Origin and Refining of Petroleum

Origin and Refining of Petroleum

A symposium co-sponsored by the Division of Petroleum Chemistry of the American Chemical Society and the Canadian Society for Chemical Engineering of the Chemical Institute of Canada at the ACS-CIC Joint Conference, Toronto, Canada, May 25–28, 1970.

H. G. McGrath and M. E. Charles,

Symposium Co-chairmen

ADVANCES IN CHEMISTRY SERIES **103**

AMERICAN CHEMICAL SOCIETY

WASHINGTON, D. C. 1971

Coden: ADCSHA

Library of Congress Catalog Card 73-164409

ISBN 8412-0120-X

PRINTED IN THE UNITED STATES OF AMERICA

FOREWORD

ADVANCES IN CHEMISTRY SERIES was founded in 1949 by the American Chemical Society as an outlet for symposia and collections of data in special areas of topical interest that could not be accommodated in the Society's journals. It provides a medium for symposia that would otherwise be fragmented, their papers distributed among several journals or not published at all. Papers are refereed critically according to ACS editorial standards and receive the careful attention and processing characteristic of ACS publications. Papers published in ADVANCES IN CHEMISTRY SERIES are original contributions not published elsewhere in whole or major part and include reports of research as well as reviews since symposia may embrace both types of presentation.

CONTENTS

PREFACE

The papers in this volume represent a useful summary of the present-day knowledge of the chemistry of petroleum in its widest sense. Beginning with a discussion of the origin of petroleum with emphasis on the chemical and migrational factors, the focus moves to properties of an important petroleum deposit—the Athabasca tar sands. The next few chapters deal with several aspects of petroleum refining, including catalytic reforming and cracking, hydrocracking, alkylation and isomerization. Developments in the petrochemical field are highlighted with emphasis on processes using ethylene, propylene, and vinyl chloride. The future of petroleum is viewed through discussions of the prospects for homogeneous catalysis and the role of oil in Canada by the year 2000.

The Canadian Kellogg Co., Ltd.
New York, N.Y.

HENRY G. MCGRATH

University of Toronto
Toronto, Ontario, Canada

M. E. CHARLES

August 1970

1

Origin of Petroleum: Chemical Constraints

GORDON W. HODGSON

Exobiology Research Group, University of Calgary, Calgary 44, Canada

The origin of oil is deeply imbedded in several dynamic environments of organic matter in sedimentary basins. Biogenic organic matter enters the system as proteins, nucleic acids, carbohydrates, and lipids. These compounds undergo (1) general degradation and polymerization and (2) general addition and subtraction of certain functional groups. The nature and extent of such changes are determined by the changing environment. Energy barriers of up to 55 kcal/mole protect many compounds. Decarboxylation of acids and reduction of alcohols are known to occur. The formation of sulfide bonds and the survival of carbon–nitrogen bonds are generally accepted. Free radical mechanisms evidently account for much of the diageneisis of organic matter. Further physicochemical control is involved in the accumulation of traces of evolving petroleum into oil fields.

Elucidation of the origin of oil, or more properly of oil fields, depends on the close interplay of interdisciplinary groups representing chemistry and geology, and it is the purpose of this paper and the accompanying paper by Hitchon (*1*) to give an overview of the entire field from these two major points of view. Obviously, each of the many subdisciplines in chemistry has a particular contribution to make, and it is the specific purpose of the present paper to draw attention to the role played by each. Principal consideration is given to the state of the art in each major area, and attempts are made to cite particular areas where intensified research efforts are required and where major advances probably will take place. No attempt is made to provide a historical survey of the field, as summarized by Hedberg (*2*), but it is interesting to note that the discipline of organic geochemistry has evolved and flourished in the last 10 years concomitantly with a sharp increase in analytical capability. While analytical capabilities result in large volumes of excellent data, in themselves they do not necessarily give rise to major conceptual advances for the origin of petroleum. The major contribution of studies in petroleum geochemistry in recent years has been the confirmation of a biogenic origin for petroleum; little has been achieved with respect to specific mechanisms for the formation of oil fields.

Analytical Chemistry

A study of the origin of petroleum requires analytical capabilities for many compounds including:

hydrocarbons
organic acids
alcohols, aldehydes, ketones, and ethers
terpenoids, including steroids and carotenoids
purines and pyrimidines
amino acids, peptides, and proteins
carbohydrates
porphyrins and chlorins
sulfur compounds
humic substances and kerogens

The methods commonly used to study these substances in the context of the chemical origin of oil relate largely to those compounds which are readily soluble in aqueous or organic solvents. For analytical purposes in general many chromatographic methods were developed. These began with early observations of Day in 1897 (*3*) on the separation of petroleum components by percolation through a column of powdered limestone. Column chromatography has been used extensively in all analytical fields, and linear elution adsorption chromatography (LEAC) for petroleum has added a new quantitative dimension (*4*). For separation and identification, paper chromatography became important for amino acids, sugars, hydrocarbons, and related substances (*5, 6, 7*). Variations on paper chromatography—*e.g.*, ion-exchange papers (*8*)—extended its usefulness, but thin layer chromatography (TLC) largely supplanted it. A number of variations of TLC broadened its applicability: continuous flow TLC for difficult resolution problems—*e.g.*, cholesterol and desmosterol (*9*); ion-exchange TLC for nucleosides using poly(ethyleneimine) hydrochloride in cellulose powders (*10*); TLC with gel permeation material for proteins (*11*). TLC methods have been developed for organic sulfur compounds (*12*). Prinzter *et al.* (*13*) presented TLC methods for a variety of mercaptans, sulfides, sulfoxides, and sulfonyl compounds. Heavy metal additives for sulfur compounds were used by Orr (*14*) for alkyl sulfides and by Murphy *et al.* (*12*) for TLC resolution of elemental sulfur, thiols, and sulfides. Chromatography in recent years, however, has depended heavily on gas–liquid methods (GC or GLC) in which the compounds to be analyzed are volatile or can be derivatized for increased volatility, as recently reviewed by Douglas (*15*). Variations on this method also include high pressure methods for coal-tar aromatics using CO_2, *n*-pentane, or 2-propanol at critical temperatures for three- to seven-ring condensed aromatics (*16*). Corwin and associates used Freon gases to separate porphyrin compounds (*17*). A major operational advance in recent years has been that of coupling a mass spectrometer to the effluent of a gas chromatograph to give direct identification of the GC peaks, as reviewed recently by Burlingame and Schnoes (*18*).

A variation on conventional adsorption chromatography is gel permeation chromatography (*19*). It is used for gross separations and size determinations, as in the case of porphyrins (*20, 21, 22, 23*). Electrophoresis was used to extend the effect of GPC separations for polypeptide chains (*24*).

Spectral methods in addition to mass spectroscopy involve standard ultraviolet, visible, infrared, nuclear magnetic resonance (NMR), and electron paramagnetic resonance (EPR); a few other spectral methods less commonly used are optical rotary dispersion (ORD) and magnetic optical rotary dispersion (MORD) as used by Stephens *et al.* (*25*) for pigments along with circular dichroism (CD) and magnetic circular dichroism (MCD). Other methods are those of luminescence, fluorescence, and phosphorescence (*26*); bioassay, as for vitamin B_{12} in organic geochemistry (*27*); activation analyses—*e.g.*, for oxygen and trace metals in asphalts (*28*); dry-column chromatography for alkaloids (*29*); extraction chromatography by elution with pH gradient (*30*); gas chromatography of optical enantiomorphs (*31*) for amino acids of ancient sediments.

The close interplay of analytical methods gives much additional information especially in those areas where the analyzed substances are very complex. Thus, while ultracentrifuge analyses for petroleum (*32, 33, 34*) gave much information, the broad approach used by others (*35, 36, 37*) involved x-ray scattering, infrared, NMR, oxidation rates, GPC, MS, pyrolysis, and vapor pressure osmometry (VPO) applied to macrostructures of asphaltic fractions. Molecular complexes substantially larger than the foregoing macrostructures are difficult to characterize by methods short of ozonolysis (*38*) and general pyrolysis (*39*).

Analyses carried out for geochemical prospecting for petroleum are commonly for light hydrocarbon gases in vegetation and soils (*40*) or in soils and unconsolidated sediments (*41*). Since such data are barely above background, they are usually subjected to statistical analysis for interpretation (*42, 43*); significant expansion of such applications were forecast on a global scale (*44*). However, a recent bibliography on geochemical prospecting (*45*) shows no entries for petroleum.

Isotope analyses for carbon, sulfur, hydrogen, and oxygen are well known. Carbon data are abundant for petroleum and associated organic matter (*46*); sulfur data are important in testing the source of sulfur in petroleum (*47*); hydrogen and oxygen isotopic data are important in defining the position of formation waters (*48*).

Application of the foregoing isolation and identification methods in the context of the origin of accumulations of petroleum permits information to be obtained at any number of significant points along the chain of events leading from the source of the organic matter to the ultimate deposit of hydrocarbons and associated compounds. Thus, analyses are carried out on:

living organisms	modern sediments
soils	sedimentary rocks
surface waters	subsurface waters
ground waters	natural asphalts

petroleum and natural gas

The results obtained from such analyses can be summarized by noting that in the general processes for forming oil fields the net effect is for hydrocarbons to increase in proportion to the total organic matter present. For example, hydrocarbons increase from about 250 ppm for typical algae (*49*) and about 35 ppm for the leaves of higher plants (*50*) to nearly 100% for petroleum. In modern sediments the data of Hunt and Jamieson (*57*) and Hunt (*52*) indicate that hydrocarbons at about 50 μgrams/gram of the total sediment constitute approximately 2500 μgrams/gram of organic matter; additional data for ancient shale sediments show that the concentration of hydrocarbons in the organic matter increases to about 15,000 μgrams/gram. Concentrations of hydrocarbons in surface waters are very low, with limited analyses giving about 0.002 μgram/gram for fresh waters and seawater (*53*). The total average bacterial biomass of seawater is estimated to be higher in some areas from the data of Kriss (*54*): 0.02 μgram/gram for the Caspian and Black Sea, and 0.0001 μgram/gram for the Pacific Ocean. The dissolved organic matter in seawater is estimated to be about 0.5 μgram/gram (*55*), from which the abundance of hydrocarbons relative to total organic matter becomes 40 μgrams/gram, not greatly different from the relative abundances in living organisms. Suspended particulate matter is commonly 0–0.05 μgram/gram (*56*) and in offshore surface samples is chiefly amorphous organic matter, at about 0.1 μgram/gram (*57*). Hydrocarbon abundances in typical subsurface waters coproduced with oil are 0.001 to 0.014 μgram/gram of water (*53*), but while data are not available for the total organic content, it may be assumed to be lower than in seawater.

Analytical data for the other classes of organic compounds show normal fatty acids to be more than an order of magnitude greater than *n*-alkanes in modern sediments, typically 26 μgrams/gram and 1.5 μgram/gram, respectively; for ancient sediments, the abundances were found to be roughly equal—*e.g.*, 32 and 43 μgrams/gram, respectively for the Mowry shale of Cretaceous age (*58*). In general, fatty acids persist into ancient sedimentary rocks. Amino acids are commonly regarded as thermally less stable but have been detected in 3.4-billion year rocks of the Precambrian by Kvenvolden *et al.* (*59*), and laboratory simulations showed stabilities to temperatures ranging from 159° to 800°C for indigenous biuret-positive substances in nautilus shells (*60*). Few data exist for other oxygen-containing compounds. Terpanes from biogenic terpenes, which are the most abundant biological hydrocarbons (*61*) appear in high concentration in petroleum (a resin concentrate of pentacyclic naphthenes was about 1% of a Nigerian crude oil) and seem to be chiefly responsible for the high specific rotations of petroleum distillates boiling around 475°C (*62*).

Purines and pyrimidines are abundant in cells, with DNA accounting for 1–3% of cellular organic carbon (*63*). Xanthine, hypoxanthine, adenine, and cytosine have been isolated as free bases from soils (*64*). Early work (*65*) showed the presence of dinucleotides in sediments, and purine nitrogen was reported for lake waters (*66*). Rosenberg (*67*) reported the detection of purine and pyrimidine bases in the hydrolysis of marine basin sediments. Ancient sediments (Devonian age) also are reported to contain purine residues (*68*). Carbohydrates are present in all living matter, and freshwater seston was found to contain about 0.4% (dry weight) of free sugars, sucrose, maltose, glucose, and fructose (*64*). Carbohydrates exist as dissolved organic matter in seawater in concentrations up to 100 μgrams/gram, as reviewed by Vallentyne (*64*). Sediments have glucose as the dominant sugar, and monosaccharides persist in Paleozoic and Precambrian fossil specimens (*68*) at levels of about 1 μgram/gram.

Chlorins are abundant in biogenic organic matter where the chlorophyll content may reach several percent of the dry weight. Oxidation to porphyrins is very slow, with only a trace being detected in modern sediments in which chlorins are commonly 100 μgrams/gram (*69*). Chlorins in lake waters were detected at about 0.0001 μgram/gram (*69*), and at much lower levels in major rivers—*e.g.*, Mackenzie river at 0.000,003 μgram/gram (*70*). Chlorins are not commonly detected in ancient sediments nor in petroleum, but porphyrins range from detection limits for such pigments in sedimentary rocks, commonly 0.001 μgram/gram, to about 10 μgrams/gram for sedimentary rocks and to 1000 μgrams/gram for petroleum (*71*). Such porphyrins invariably show a wide range of molecular sizes (*72*).

Sulfur compounds are abundant in petroleum, but one of the major gaps in the analytical chain between source and end product lies in this field. Thin layer chromatography is useful for isolating and identifying sulfur and organic sulfur compounds such as thiols and sulfides—*e.g.*, 5-octadecyl thiol and 4-di-*n*-octadecyl trisulfide (*73*). Simple thiols, sulfides, alkyl sulfides, and disulfides, certain cyclic sulfides, sulfoxides, and isothiocyanates are present in plants (*74*). Mercaptans were studied by Adams and Richards (*75*) in an anoxic fjord. Free sulfur is plentiful in modern sediments—*e.g.*, 7.5% sulfur with 2.1% organic carbon in an algae muck sample (*76*). Proteins are the principal organic source of sulfur and might be expected to contribute up to 1% sulfur content of the total organic matter (*77*). Sulfur compounds in petroleum as summarized by Speers and Whitehead (*78*) are primarily thiols in boiling ranges below 150°C. In distillates up to 250°C thiacycloalkanes, bi- and tricyclothiaalkanes and thiophenes predominate, and in still higher boiling ranges benzothiophenes become important. Compounds with two sulfur atoms are uncommon.

The analysis for humic materials and kerogens is much less specific than for individual compounds as in the foregoing instances, but the amounts involved are enormous, with 10 times as much humic acid material existing on earth as there is organic matter in living organisms, according to a recent review by Stevenson and Butler (*79*). Humic substances are normally recovered by extraction with 0.5*N* NaOH and are subdivided into humic acid, fulvic acid, hymatomelanic acid, and humin on the basis of solubility characteristics. Functional group analyses give carboxyl, phenolic, and enolic hydroxylquinone, hydroxyquinone, lactone, ether, and alcoholic hydroxyl groups. Humic substances may exist in trace quantities in plants but are abundant in sediments—*e.g.*, marine modern sediments with about 30% humic material (*80*) and in soils. Ground waters are reported to contain humic acids along with naphthenic acids (*81*) as reported by Davis (*77*). Fresh waters contain high concentrations of humic material, and Corbett (*82*) estimated that 5% of the dissolved solids entering the oceans were humic materials and would account for nearly 100 million tons per year entering the marine sediments. Ancient sedimentary shales contain on the average 20,000 μgrams/gram of kerogen, roughly the same as the humic material in modern sediments (*52*). The analysis of humic materials and kerogens by pyrolysis methods is finding favor, as illustrated by Douglas *et al.* (*39*) in which Green River shale was heated at 500°C to give yields of alkanes, alkenes, and aromatic hydrocarbons greater by a factor of 10 when compared with raw extracts of the same material.

The foregoing incomplete but illustrative outline of the role of analytical chemistry in the origin of petroleum makes it possible to discuss the position of a number of the other subdisciplines in connection with the genesis of oil fields.

Biochemical Aspects of the Origin of Petroleum

Biological agents have two major functions in the question of the origin of oil: biogenesis—the initial generation of organic matter—and biodiagenesis—the alteration of organic detritus.

Biogenesis. Hydrocarbons are synthesized by most organisms. The mass production is quite substantial—6×10^7 barrels of hydrocarbons per year (*77*)—even though concentration of hydrocarbons in the biomass is small. Mechanisms for the complete biosynthesis of such compounds are not clearly known, but almost certainly many hydrocarbons are derived from biosynthetic fatty acids (*83, 84*). The initial generation of fatty acids is believed to occur through the addition of C_2 units to other fatty acid units with the intervention of coenzyme A and DPNH. In this manner, fatty acids build up in units of two and give rise to even-carbon numbered molecules. Branched fatty acids are erected in a similar way with branched amino acids involved in the process (*85*). Unsaturated acids, including acetylenic compounds (*86*), are also biosynthesized. Amino acids are synthesized biologically through the reaction of ammonia with α-ketoglutarate acid to make glutamine; this is a particular expression of generating amino acids from corresponding keto acids through transamination (*87*). Nucleic acids are synthesized starting with the biosynthesis of purines, and nearly all living organisms are able to carry out this kind of synthesis. Involved in the synthesis are formate, CO_2, glycine, aspartic acid, and glutamine. Biogenesis of porphyrins found in petroleum is essentially that of protoporphyrin IX. As reviewed by Lascelles (*88*) this starts with the condensation of eight glycine molecules with eight succinyl-coenzyme A units to give δ-aminolevulinic acid which dimerizes readily to porphobilinogen. Decarboxylation and mild oxidation in a number of steps produce protoporphyrin from which hemes and chlorophylls arise.

Isoprenoid structures for carotenoids, phytol, and other terpenes start biosynthetically from acetyl coenzyme A (*89*) with successive additions giving mevalonate, isopentyl pyrophosphate, geranyl pyrophosphate, farnesyl pyrophosphate (from which squalene and steroids arise), with further build-up to geranyl geranyl pyrophosphate, ultimately to α- and β-carotenes, lutein, and violaxanthin and related compounds. Aromatic hydrocarbon nuclei are biosynthesized in many instances by the shikimic acid pathway (*90*). More complex polycyclic aromatic compounds are synthesized by other pathways in which naphthalene dimerization is an important step (*91*).

Biodiagenesis of Organic Matter. Two aspects of this type are important: one, in which organic matter is biotically modified to produce constituents of petroleum, and the other, in which petroleum is modified and even destroyed by the action of microorganisms. While both kinds of processes may take place concurrently, they are more readily treated separately.

BIODIAGENESIS IN THE FORMATION OF CONSTITUENTS OF PETROLEUM. The basic question is to what extent are the components of petroleum generated by microorganisms? In general, the aerobic microbes in upper sediments are the agents that create anaerobic conditions by removing oxygen from the environment (*77*). The role of microorganisms thus includes the creation of anaerobic conditions which markedly slow down the microbial decomposition of petroleum-source material; secondly, the production of organic material which more closely resembles petroleum by lowering the oxygen, nitrogen, sulfur content of the organic constituents of depositing sediments. Deamination and decarboxylation are attributed to anaerobic microbial activity in the reducing strata of modern sediments (*92*). Microbial anaerobic metabolism of lipids has two effects: saponification of the esters and saturating the unsaturated, (*e.g.*, linoleic acids (*93*)). Denitrifying bacteria are important in this area (*94*) as are sulfate-reducing bacteria. Recent sediments appear to contain appreciable quantities of hydrogen-producing microorganisms (*77*), and the hydrogen is evidently involved in the formation of methane from both CO_2 and CO (*95*). Hydrogen also is utilized in the reduction of organic compounds—*e.g.*, fumarate by *Esherichia coli* (*96*). Suspensions of *Clostridium sporogenes* in the presence of molecular hydrogen readily deaminate amino acids and alter other compounds (*97*). Hydrocarbons are formed by sulfate-reducing bacteria from marine sediments when grown on CO_2 and H_2 in a mineral salts medium (*98*). Phytol is converted to pristane by zooplanktonic copepods (*99*), and this kind of reaction may be important in explaining the presence of isoprenoid hydrocarbons commonly found in living organisms (*100*).

BIODIAGENESIS IN THE DESTRUCTION OF HYDROCARBONS. Davis (*77*) reported that all samples of sediments in an API Project 43A study of marine sediments regardless of location, water depth, or core depth contained hydrocarbon-oxidizing bacteria, but the oxidation was inhibited by the presence of H_2S above 0.001 mole/liter. Soils are also well known for the presence of microorganisms capable of oxidizing hydrocarbons—*e.g.*, the *Pseudomonas* species. The relationship between inhibition by H_2S and the presence of hydrocarbons was related (*77*) as a possible explanation for richest hydrocarbon accumulations to be commonly found under H_2S "protection." Winters and Williams (*101*) reported that about 10% of all crude oils examined had compositional characteristics strongly suggestive of microbial alteration. Microbial oxidation of aromatic hydrocarbons is possible (*102*), but the benzene ring remains intact while alkyl side chains are oxidized. *Desulfovibrio desulfuricans* isolated from a freshwater oil-bearing aquifer slowly oxidized methane, ethane, and *n*-octadecane, and it is suggested that the bacteria may generate hydrogen sulfide in subsurface formations by coupling the oxidation of hydrocarbons to sulfate in reduction (*103*). Microflora including sulfate reducers were reported for oil accumulations in the U.S.S.R., and laboratory simulations of microbiological attack showed a general breakdown of the resinous components and the evolution of free nitrogen (*104*).

Inorganic Chemistry

Chemical aspects of the origin of oil in the sense of inorganic chemistry relate largely to the interaction of inorganic substances such as mineral surfaces and dissolved ions with biogenic organic matter. The interactions involve both chemical and physical interplays but have little to do with largely discredited theories on the inorganic origin of petroleum.

Mineral Surfaces. Organic matter is chemically adsorbed (derivatized) at the surfaces of clay minerals, zeolites, and related minerals (*105*) and is at times protected, concentrated, and degraded by contact with the solid surfaces. For example, porphyrins are protected (*106*), as are optically active amino acids by montmorillonite (*107*). This may result in part from the position of the organic matter in lattice spaces, as shown by Stevenson and Cheng (*108*) for proteinaceous substances keyed into hexagonal holes on interlamellar surfaces of expanding lattice clays, or from the fact that there are ordered structures at solid–water interfaces (*109*).

Soluble Ions. Organic matter generated by living organisms from carbon dioxide and water is at most stages in its existence in contact with aqueous systems replete with the common cations and anions and also with a wide variety of trace ions. Some are uniquely complexed in specific chemical structures, as in the case of chlorophylls and hemoglobins, and as such are part of the biotic system. Others may be regarded as being simply present in the same microsystem, and when the living processes cease, they tend to form simple complexes with the degrading organic matter. An example is the chelation of trace metal ions with amino acids (*110*). Similar complexes evidently arise between metals and sugars and also with evolving humic substances (*111*) as well as with petroleum asphaltenes (*112*). The relationship between inorganic ions and humic material was examined in connection with geochemical anomalies involving lead, zinc, copper, nickel, and cobalt (*113*). Opportunities exist for initial contact of organic matter in rivers with dissolved ions (*114*). Interaction between metallic ions with both dissolved and humic material was recently examined by Koshy *et al.* (*115*,

116). The role of hydrothermal waters in generating petroleum from organic polymers was discussed by Mueller (*117*). Whether any specific hydrocarbon-forming reactions occur as a result of such complexing processes is not known at this time. Other organic constituents react with associated inorganic ions—*e.g.*, porphyrins almost certainly form chelates with nickel and iron in this way and possibly with vanadium as well (*69*). Few experiments have been done to demonstrate the interaction of inorganic ions with organic substances at this stage of the genesis of petroleum.

Sulfur. Undoubtedly, many of the reactions involve or are influenced by the presence of sulfur in the petroleum-forming stages as they are in the formation of mineral accumulations (*118*). Indeed this critical stage of the formation of petroleum is profoundly influenced by the redox conditions established by the inorganic sulfur system.

$$SO_4^{2-} \rightleftarrows S^0 \rightleftarrows H_2S \rightleftarrows HS^- \rightleftarrows S^{2-}$$

In the early stages in the diagenesis of organic matter much of the sulfur that ultimately remains in the organic system is introduced. It probably enters through interaction with sulfur in the elemental form giving rise to sulfides and mercaptans. Organosulfur compounds in turn affect the stability of saturated hydrocarbons, with benzenethiol and *tert*-butyl disulfide accelerating the thermal decomposition of branched chain alkanes (*119*). Elemental sulfur reacts readily with cholesterol at 150° and with farnesol at 135°C, producing one-, two- and three-ring aromatic hydrocarbons (*120*). A somewhat artificial simulation of the effect of sulfur in natural asphalts is afforded by the work of Tucker and Schweyer (*121*) on the distribution and reactions of sulfur in asphalt during air blowing and sulfurizing reactions. Nevertheless, polysulfide linkages readily formed at low temperatures and evidently resulted in cyclic sulfides and thiophenes. Toland (*122*) demonstrated the oxidation of alkylbenzenes with sulfur and water at 200°–400°C.

Inorganic Aspects of Hydrocarbon Transport. Trace hydrocarbons existing in unconsolidated and consolidated sediments are subject to mobilization and transportation in aqueous systems, and the efficiency of such processes is undoubtedly related to the presence of dissolved salts. Hydrocarbons in the C_{11}–C_{19} range are accommodated less easily in saline solutions than in freshwater solutions by a factor of about 5 for salinities of 1000 μgrams/gram (*123*). On migration through an unconsolidated sand bed, aqueous systems carrying hydrocarbons were observed to be very stable when the salinity arose from $NaHCO_3$ but unstable when the salt was NaCl, at the same concentrations. Accommodation of hydrocarbons was affected more severely in the presence of calcium salts. As a result, the occurrence of various inorganic salts at various levels of concentration in subsurface water (*124*) must have a profound effect on the mobilization, movement, and deposition of hydrocarbons.

As part of the over-all topic of the role of inorganic ions in the formation of economic accumulations of hydrocarbons—*i.e.*, oil-fields—the analogy with the formation of ore bodies stands out clearly. The closeness of the analogy is revealed in a number of instances where hydrocarbon accumulations actually accompany the formation of mineral deposits. Such occurrences are hydrothermal in character (*117*), and while the oil deposits are trivial in amount, the association is very real, being most commonly observed in the case of ore deposits involving lead, zinc, and uranium.

Inorganic Origin of Petroleum. Suggestions are made from time to time that hydrocarbons arise from nonbiogenic precursors. While this may be so in the cosmic sense for complex organic matter, there seems to be no significant body of evidence to indicate that any appreciable proportion of the terrestrial oil accumulations arises from anything other than from biogenic sources (*125*).

Organic Chemistry

The formation of hydrocarbons from organic matter in general can be visualized through a number of key reactions involving removal or modification of functional groups. Major reactions are those involving reduction and decarboxylation. In addition, some reactions involve fragmentation of polymeric material and reactions in which sulfur is important. A novel approach was used by Yen and Silverman (*126*) in which petroleum maturation and other petroleum transformations were evaluated by parameters directly related to the structure of asphaltenes and the diversity of homologs of porphyrins occurring in petroleum.

Reduction. Many organic compounds produced by organisms are oxygenated to the extent that they are alcohols (including polyols such as sugars), aldehydes, ketones, and acids, for example fatty acids. Complete reduction of such compounds, of course, produces hydrocarbons. Other biogenic substances are unsaturated—*e.g.*, carotenes— and reduction of these compounds also results in saturated hydrocarbons. Alcohols are difficult to reduce, however, as indicated by the observation that the hydroxyl groups of most alcohols can seldom be cleaved by catalytic hydrogenation, with only benzyl type alcohols readily undergoing the reaction with palladium-on-charcoal as a catalyst. Aldehydes and ketones are uncommonly isolated from reductions of acids, the reduction normally proceeding directly from acid to alcohol. Ethers are commonly cleaved by hydrogenation with alcohols as intermediates, and this reaction has relevance to the breakdown of polymeric materials such as kerogens and lignins. The difficulty with the direct reactions involving reduction of organic compounds is that alcohols must be reduced under conditions which are plausible geochemically. Such conditions are not readily recognized, and hence suggestions that the bulk of hydrocarbons arise in this manner cannot be readily accepted in terms of reactions of classical organic chemistry. It must be recalled, however, that such reactions may take place in low yield to the extent that significant quantities of hydrocarbon product are generated. In addition, it is obvious that reductions of alcohols actually take place in the origin of petroleum—*e.g.*, in the case of sterols yielding steranes (*78*)—but the mechanism of the reaction is not known; it may depend on microbial activity. In the same connection, the abundance of glycerol in biogenic material would be expected to give rise by hydroreduction to abnormal amounts of propane. Although reduction of phytol to phytane in sediments was regarded as a source for a major petroleum isoprenoid (*127*), this kind of reaction is evidently not required in the light of the existence of isoprenoid alkanes in living organisms (*99, 128*). Another mechanism for the removal of alcohol hydroxy groups is through dehydration reactions as demonstrated by Andreev *et al.* (*129*). Simulated reduction of humic substances using zinc reduction was successful in producing condensed polycyclic substances—*e.g.*, perylenes (*130*).

Decarboxylation. A second major potential pathway for the formation of hydrocarbons from oxygen-containing organic matter involves decarboxylation of organic acids. The possibility for major contributions from this pathway seems much higher than with alcohols, as indicated by field observations and laboratory simulations. The field observations relate primarily to the relationship between *n*-fatty acids and *n*-alkanes (*58*) and to that between carboxylated porphyrins and decarboxylated porphyrins (*69*), from which considerable evidence is available to show that conversions of the acid precursors take place *in situ*. Laboratory studies confirm the indicated changes, with *n*-fatty acids (*e.g.*, C_{22}) being converted in wet (and dry) clay-containing systems under relatively mild thermal conditions (*131*). In addition to the direct decarboxylation reaction, other processes ran concurrently, and these produced appreciable quantities of *n*-alkanes of both shorter and longer chain length. The significance of this observation lies pragmatically in the fact that it helps to explain the loss of odd-carbon dominance in petroleum, and mechanistically in the fact that free-radical mechanisms are almost certainly operative in mild geochemical environments (*132, 133*).

A further pathway for the formation of specific individual hydrocarbons is that of the chemical degradation of amino acids. The reactions may be summarized as follows:

$$R—CH\,NH_2—COOH + 2H \rightarrow NH_3 + RCH_2COOH \quad (1)$$

$$RCH_2COOH + 8H \rightarrow RCH_3 + CH_4 + 2H_2O \quad (2)$$

$$RCH_2COOH \rightarrow RCH_3 + CO_2. \quad (3)$$

Reductive deamination of amines takes place through R—N=N—H intermediates, with the involvement of carbonium ions. Primary amines may be reduced by HNF_2, but the geochemical plausibility of such a reaction or that for quaternary ammonium salts using lithium aluminum hydride is very difficult to estimate. Reactions 2 and 3 are simply those for regular unesterified fatty acids described above, with the further observation that the resulting hydrocarbons are primarily of short chain length, thus giving a kind of rational origin for low molecular weight alkanes of petroleum source material (*134*).

Spallation. While the foregoing reactions dealt with the generation of specific hydrocarbons from specific chemical precursors, a significant body of alkanes undoubtedly arises from the more or less random fragmentation of polymeric organic matter. Thus the ill-defined organic polymers of direct biotic origin, those of initial polymerization in organic detritus, and those of long standing in sediments and sedimentary rocks give rise to fragments which in many instances are alkanes. This area of study has profited from the study of carbon isotopes (*135*), from which it is clear that spallation of alkanes from complex organic matter takes place, as inferred from the isotope examination of petroleum fractions. Laboratory simulations of mild pyrolysis of kerogens give a variety of products commonly containing significant quantities of unsaturated hydrocarbons as well as normal and branched alkanes (*39*). These are simple thermally generated fragments resulting from the random breaking of carbon–carbon bonds. In addition, other bonds must be involved, probably C–S, C–N, and C–O, giving rise to the corresponding free radicals for subsequent repolymerization to alkanes and new polymers, perhaps ultimately through asphaltenes to graphitic material. In general, the effect of time and depth of burial on the naphtha as gas oil content of crude oils shows an increase in naphtha and a decrease in gas oil (*136*), and data for the same phenomenon were presented (*137*) for hydrocarbons extracted from ancient sediments.

Sulfur Compounds. Sulfur surely plays a role in the organic chemistry of the origin of petroleum. Many of the direct reactions were considered in foregoing sections—*e.g.*, the aromatization of unsaturated hydrocarbons through interaction with elemental sulfur (*100*). In considering the genesis of hydrocarbons from sulfur-containing compounds reactions such as the following can be considered for sulfur-containing amino acids, cysteine, cystine, methionine, and penicillamine.

$$HS{-}CHNH_2{-}COOH + 2H \rightarrow CH_3\,NH_2COOH + H_2S \quad (1)$$

$$CH_3{-}S{-}(CH_2)_2{-}CHNH_2{-}COOH + 4H \rightarrow CH_4 + H_2S + CH_3CH_2CHNH_2{-}COOH \quad (2)$$

Such reductive desulfurizations for hydrogenolysis are readily carried out in the laboratory using Raney nickel; other catalysts—*e.g.*, molybdenum sulfide, zinc sulfide—do not poison as readily and are commonly used for commercial desulfurizations. Again the relevance of this kind of organic chemistry to geochemical situations is not clearly evident. However, if such reactions were possible, they would be helpful in accounting not only for low molecular weight hydrocarbons as in the case of amino acids but also possibly for the formation of H_2S as a constituent of natural gas, although much of the H_2S is probably derived from microbial reduction of elemental sulfur.

Methane. Considerable attention must be given to the origin of methane in terms of organic chemical reactions. The abundance of this compound relative to liquid hydrocarbons is substantial, commonly amounting in mass to more than the liquid hydrocarbons in a given basin. Methane can be formed as follows:

$$RCH_3 \rightarrow R\cdot + CH_3\cdot \quad (a)\ (1)$$

$$CH_3\cdot + H\cdot \rightarrow CH_4 \quad (2)$$

$$CH_3\cdot + CH_3\cdot \rightarrow C_2H_6 \quad (3)$$

$$H_2N{-}CH_2{-}COOH + 2H \rightarrow NH_3 + CH_3{-}COOH \quad (b)\ (1)$$

$$CH_3{-}COOH \rightarrow CH_4 + CO_2 \quad (2)$$

The free radical reactions given in *a* refer to a variety of processes, particularly spallation reactions, whereas those of *b* are specific. While the spallation source for methane is clearly dominant over that from glycine, it is difficult to judge whether it also grossly exceeds that for the production of methane by microorganisms.

Physical Chemistry

The chemical constraints related to physical chemistry in the origin of petroleum should be more clearly defined than in the other subdisciplines, but the complexity of the situation becomes overwhelming. Physicochemical aspects are widespread, fitting into two major areas: one dealing with reaction kinetics and thermodynamics, and the other with physical relationships including colloid chemistry.

Thermodynamics. The role of chemical thermodynamics in the origin of petroleum is concerned largely with demonstrating an almost self-evident drive from unstable biogenic molecules to stable hydrocarbons and related compounds of petroleum. Dayhoff *et al.* (*138*) made extensive computer studies of the thermodynamic interplay in systems comprising carbon, hydrogen, and oxygen. The systems considered were for ideal gases at various temperatures and pressures, and the results were computed from known free energies of formation and presented in the form of ternary diagrams. Two pronounced thresholds appeared, at which the concentrations of many compounds changed abruptly by many orders of magnitude. The first was the oxygen threshold along the line joining CO_2 and H_2O. Passing from below this line, from the oxygenated region toward the carbon apex, traces of many organics appear. The next threshold is an "asphalt" threshold, above which large concentrations of benzene and polycyclic aromatics appear. While the position of this threshold is a function of temperature and pressure, at 500°K and 1 atm pressure, the asphalt threshold lies along the CH_4–CO_2 line. To broaden the study and increase its relevance to the origin of complex organic matter, nitrogen and sulfur were introduced to the hypothetical systems in later phases of the study.

Andreev *et al.* (*129*) reviewed the position of thermodynamics in the transformation of petroleum in nature. For most conversions of organic matter related to the formation of petroleum the value for changes in free energy are:

$$-10 \text{ kcal/mole} < \Delta F < +10 \text{ kcal/mole}$$

Isomerization processes clearly illustrate thermodynamic relationships. For example, for the isomeric hexanes the highest level of free energy (relative to standard states) is that of *n*-hexane (−1.04 kcal/mole) while isomeric 2-methylpentane is −1.95 kcal/mole. Further branching gives −2.82 kcal/mole for 2,2-dimethylbutane. Similar relationships hold for cyclic hydrocarbons, and the same energy drives account for disproportionation and hydrogenation:

$$\underset{\displaystyle CH_3}{CH_3{-}\overset{}{\underset{|}{CH}}{-}CH_2{-}CH_3} + H_2 \rightarrow CH_3{-}CH_2{-}CH_2{-}CH_3 + CH_4$$

for which $\Delta F = -12.39$ kcal/mole (*129*, p. 168).

The conversion of a number of typical hydrocarbons was considered (*129*, p. 178). For example, under equilibrium conditions with an adequate supply of free hydrogen, benzene is converted to cyclohexane and hexane

$$C_6H_6 + 3H_2 \rightarrow C_6H_{12} \quad \Delta F^\circ_{298^\circ} = -23.67 \text{ kcal/mole}$$

$$C_6H_{12} + H_2 \rightarrow C_6H_{14} \quad \Delta F^\circ_{298^\circ} = -6.95 \text{ kcal/mole}$$

It is interesting to compare the foregoing data with that for interaction with heteroatoms:

$$C_6H_{14} + \frac{1}{2}O_2 \rightarrow C_6H_{12} + H_2O$$

for which $\Delta F°_{298°} = -49.74$ kcal/mole. The corresponding theoretical reaction with sulfur has a free energy change of −1.35 kcal/mole.

REACTION RATES. Thermodynamic data are of primary importance in defining the direction of reaction, and complementary data for rates of reaction are required to determine whether the reaction will take place to any significant extent. Reaction rates in turn are influenced greatly by catalysis. In connection with the origin of oil the catalytic factor is difficult to ascertain because the reaction systems are characterized by low temperatures and the presence of water, for which few data are available. As reviewed by Eisma and Jurg (*132*) the cracking of a paraffin can be subdivided into thermal cracking involving a radical mechanism and catalytic cracking with a carbonium ion mechanism. Free radical mechanisms were postulated for the decomposition of behenic acid, yielding not only the C_{21} alkane but other alkanes from C_{16} to C_{28}. Reaction rates were such that appreciable yields were obtained after about 300 hours at about 250°C. While the foregoing reaction was essentially a thermal catalytic decarboxylation process, it was substantially different from a thermal degradation process for heterocyclic aromatic compounds—a porphyrin fraction from petroleum—in which higher temperatures were required to produce comparable rates of reaction. The Arrhenius plots gave sharply differing superficial activation energies—*viz.*, about 10–30 kcal/mole (increasing with temperature) for the decarboxylation and about 52.5 kcal/mole for the thermal degradation (*139*). The significance of the indicated thermal degradation of porphyrins, confirmed by Rosscup and Bowman (*140*), in geochemical situations is of marginal magnitude, but the much higher rates for the conversion of fatty acids appear to be well within the realm of geochemical plausibility. It is reasonable, therefore, to postulate that other organic conversions are likely to take place as well even though by conventional chemical concepts they are "impossible." Thus, experiments by Vallentyne (*141, 142*) indicated that amino acids could be decarboxylated under geochemical conditions. Reaction rates for the thermal degradation of four amino acids in oxygen-free water at 0.01*M* concentration were found to be: pyroglutamic acid, $k = 2 \times 10^9\ e^{-35,800/RT}$; phenylalanine, $k = 2 \times 10^8\ e^{-30,800/RT}$; threonine, $k = 2 \times 10^{12}\ e^{-33,800/RT}$; and serine $k = 4 \times 10^9\ e^{-29,350/RT}$ (*141*). For typical geochemical situations, these reaction rates indicate that massive degradation of amino acids takes place within short periods of time, yet amino acids are found in sediments of Precambrian age (*143, 144*). It is possible however, that some of the simple organic molecules entered the sediments since then (*145*). Thermal transformations leading to the formation of hydrocarbons appear to occur under relatively mild conditions to the extent that hydrocarbon generation in some localities is now evidently complete (*146*).

HETEROGENEOUS *vs.* HOMOGENEOUS REACTIONS. The catalytic role of clay minerals in the origin of petroleum (*147*) has not been studied exhaustively, but Weiss (*105*) reviewed illustrative reactions involving amino acids. Mitterer (*148*) examined the degradation of amino acids in the presence of carbonates and found what appeared to be catalysis. Metal complexing of chlorins with nickel ions was found to occur readily in homogeneous solution over the temperature range 74°–115°C (*149*), but little increase in reaction rate if any was noted for the corresponding complexing of the metal in the presence of clay minerals (*150*). Thermal studies on the degradation of carbohydrate residues in Paleozoic shales showed that the clay minerals of the shale acted as a protective medium up to about 200°C, as reported by Swain (*151*). It is possible that many of the potentially active sites on silica are modified by organic interaction (*152*). Earlier suggestions that clay surfaces act as acid catalysts seem unacceptable in the presence of water which appears to prevent carbonium ion reactions (*132*). Others attributed some particular reactions to pH effects, in which the dispersed minerals acted as pH buffers—*e.g.*, in the primary degradation of chlorophyll (*153*).

The geochemical environment is clearly a heterogeneous reaction system, comprising as it does water, minerals, and organic matter. Some of the organic matter is soluble in the water, much of it is not. At the outset there is essentially no continuous liquid organic phase in which homogeneous reactions would be expected to take place. Obviously many of the chemical conversions occur at interfaces, but the possibility that microsystems exist for homogeneous reactions cannot be ruled out. This is true, for example, for the thermal degradation of soluble organic matter such as amino acids and sugars.

PHYSICAL AND COLLOIDAL ASPECTS. Much of the effort of geochemical research to date has gone into the foregoing aspects of the origin of oil, and little attention has been given to the mechanisms by which the components of petroleum are formed into an oil field. Hydrocarbons are clearly ubiquitous (*52*); the fact that they are transported is equally clear. How they are transported is not clear. Consider a C_{22} fatty acid decarboxylated in an aqueous system containing clay minerals and a complex agglomeration of organic matter. The fatty acid is not very soluble in the water; it may be weakly adsorbed on a montmorillonite surface or imbedded within the organic matrix. On decarboxylation, it becomes even less soluble in the aqueous medium. It becomes less strongly adsorbed on the clay mineral and perhaps less deeply buried in the organic mass. The water is moving, but imperceptibly, and contains dissolved salts. The concentration of hydrocarbons in the sediment is perhaps between 1 and 10 μgrams/gram in terms of the water content. The solubility of the C_{21} alkane is less than 10^{-6} μgram/gram, yet C_{21} alkanes can be dispersed mechanically to levels well above this, evidently as stable colloids, even in solutions of dissolved salts—*e.g.*, $NaHCO_3$ (*123*). Other organic matter is present, and some of it surely is in true solution. Some may be present as surfactants. Hydrocarbons are present in surface waters of lakes to about 0.0005 μgram/gram in seawaters to 0.0002 μgram/gram, in oil-field waters to 0.014 μgram/gram (*53*). Waters expressed from compacting nonmarine sediments were observed to contain about 0.014 μgram/gram, perhaps significantly higher than those for some oil-field waters. Increasing salt contents sharply reduce the capability of water to accommodate dispersed hydrocarbons, and divalent cations such as Ca^{2+} are much more destructive than Na^{+} on the negatively charged colloids (*123*). Similar experiments were done by Dickey (*154*) using crude oil, tap water, and differentially packed porous flow channels.

McAuliffe (*155, 156, 157*) examined the solubility of hydrocarbons in water, and Franks (*158*) in reviewing the results concluded that long chain hydrocarbons are vastly more soluble than would be expected from an extrapolation of the standard free energy change of the lower members of the series. These results suggested that substantial modifications in the structure of the water are produced by the presence of the solutes; this was developed in some detail by Baker (*159*) following his earlier work on the role of surfactants (*160, 161, 162, 163*). The importance of surfactants was explored in the earlier literature in terms of production practice (*164*) and of the presence of porphyrins and other nitrogen-containing compounds (*165*). The reverse surfactant position of water in oil was recently examined (*166*). The relationship of naphthenic acids as natural surfactants to the accumulation of oil in an artesian aquifer was examined by Davis (*167*), and the remote possibility that hydrocarbons and organic matter form crown complexes with inorganic salts (*168*) should not be overlooked. The effect of soluble gases in formation waters was studied by Sokolov *et al.* (*169*), who found that carbon dioxide enhanced the solubility of organic substances in water. The same topic was examined by Meinschein *et al.* (*170*), and comprehensive studies of the solubility of gases in synthetic brines were carried out by O'Sullivan *et al.* (*171*). A syngenetic relationship between high wax oils and terrigenous organic matter was noted by Hedberg (*172*).

The general movement of formation fluids is of paramount importance in the transport of hydrocarbons (*173, 174*). In a companion paper to the present study Hitchon (*1*) related in considerable detail the position of basin-wide movements of formation fluids based on a number of earlier studies (*48, 124, 175, 176, 177*). Brezgunov *et al.* (*178*) also used isotope studies to explain the origin of ground waters and petroleum. A classical approach to the migration of the movement of oil and gas was emphasized by Silverman (*179*) in terms of source rock and primary and secondary migration. That changes in oil composition should take place is obvious (*180*, p. 212). Attempts have been made to relate fluid composition whether colloidal or not (*38*) to natural chromatographic processes (*181, 182, 183, 184*). The theory of deep filtration in the flow of suspension through porous media as recently reviewed by Herzig *et al.* (*185*) is of considerable relevance in predicting the movements of hydrocarbons suspended in water. The anisotropic permeability (*186*) and physics and thermodynamics of capillary action in porous media (*187*) similarly are of fundamental importance. At shore lines, the mixing of ground water and seawater is critical in determining the organic load of the water (*188*), and the same is true for the contact of ground waters with subsurface waters in petroleum provinces (*189*). The analogy with the modification of the accommodation of minerals in water is close—*e.g.*, in the case where mineral-bearing ground waters undergo sharp changes in pH and P_{CO_2} (*190*). The diffusion of hydrocarbons in waters was recently considered by Jeffrey and Zarrella (*191*) and by Bonoli and Witherspoon (*192*), who found substantial salt effects related to structure breakers. The movement of formation fluids may be indicated by the escape of volatile constituents. Thus, geochemical exploration for petroleum is currently based on the upward diffusion of light hydrocarbon gas (*41, 193, 194*).

Discussion

The foregoing sections give illustrative sketches of several aspects of the chemical origin of oil. The picture that emerges is that petroleum hydrocarbons are generated biogenetically or are derived from biogenic substances by low yield reactions of uncertain mechanism commonly involving biotic agencies as well as low temperature catalytic reactions involving free radical conversions. The formation of oil fields is a much larger and more important topic; it depends largely upon the principles of physical and colloid chemistry to explain the mechanisms of the mobilization, transportation, and accumulations of hydrocarbons while continuous alterations are taking place in the organic constituents of the sedimentary basin as a result of chemical and thermal treatment.

As noted earlier (*173*) the effort expended on defining the source of petroleum hydrocarbons has not resulted in a complete understanding of the origin of oil fields nor has it helped greatly in exploring for petroleum.

The chemical constraints evolving out of the present study appear to be almost minimal in scope. Hydrocarbons are generated from a variety of compounds of biogenic origin. The chemical conditions required are not strikingly restrictive. Conditions appear to be favorable for the presence of hydrocarbons nearly everywhere. In some instances, the processes of formation may have passed their peak, but it seems difficult to believe that there are areas in which absolutely no hydrocarbons exist. Obviously, however, there are vast areas in which oil fields do not occur. It seems probable that this is not caused by a failure to generate hydrocarbons but rather by a failure to have a coincidence of all of the required factors for oil field formation involving mobilization, transportation, and accumulation of hydrocarbons. Physicochemical factors are of vital importance in all such areas, and while these can be stated in terms of entropy states in thermodynamics, the more productive approach may be to postulate and test the physicochemical mechanisms involved. For example, it may be possible to show in a particular area that the migrating subsurface waters are completely unfavorable for the mobilization of dispersed hydrocarbons, and consequently the search for oil accumulations should be focussed where mobilization conditions are favorable and there is a coupling with geochemical factors that promote the disaccommodation of hydrocarbons from migrating subsurface waters. This, in direct concert with the geological constraints, will certainly lead to higher success ratios in oil field exploration.

Acknowledgment

This study was supported by a grant from the Research Council of Alberta.

Literature Cited

(1) Hitchon, B., "Origin of Oil: Geological and Geochemical Constraints," ADVAN. CHEM. SER. (1971) **103**, 30.
(2) Hedberg, H. D., "Geologic Aspects of Origin of Petroleum," *Bull. Am. Assoc. Petrol. Geol.* (1964) **48**, 1755.
(3) Tiseluis, A., "Some Recent Advances in Chromatography," *Endeavour* (1952) **11**, 5.
(4) Snyder, L. R., "Linear Elution Adsorption Chromatography," *Advan. Anal. Chem. Instrumentation* (1964) **3**, 251.
(5) Rittenberg, S. C., Emergy, K. O., Hulsenann, J., Degens, E. T., Fay, R. C., Reuter, J. H., Grady, J. R., Richardson, S. H., Bray, E. E., "Biogeochemistry of Sediments in Experimental Mohole," *J. Sediment. Petrol.* (1963) **33**, 140.
(6) Degens, E. T., Emergy, K. O., Reuter, J. H., "Organic Materials in Recent and Ancient Sediments. Part III: Biochemical Compounds in San Diego Trough, California," *Neues Jahrb. Geol. Paleontol. Monatsh.* **1963**, 231.
(7) Hoffmann, D., Wynder, E. C., "On the Isolation and Identification of Polycyclic Aromatic Hydrocarbons," *Cancer* (1960) **13**, 1062.

(8) Sherma, J., Strain, H. H., "Chromatography of Leaf Chloroplast Pigments on Ion-Exchange Papers," *Anal. Chim. Acta* (1968) **40,** 155.
(9) Neuhard, J., Randerath, E., Randerath, K., "Ion-Exchange Thin-Layer Chromatography," *Anal. Biochem.* (1965) **13,** 211.
(10) Lees, T. M., Lynch, M. J., Mosher, F. R., "Continuous Flow Thin-Layer Chromatography," *J. Chromatog.* (1965) **18,** 595.
(11) Morris, C. J. O. R., "Thin-Layer Chromatography of Proteins in Sephadex G-100 and G-200," *J. Chromatog.* (1964) **16,** 167.
(12) Murphy, M. T. J., Nagy, B., Rouser, G., Kritchevsky, G., "Identification of Elementary Sulfur and Sulfur Compounds in Lipid Extracts by Thin-Layer Chromatography," *J. Am. Oil Chemists Soc.* (1965) **42,** 475.
(13) Prinzler, H. W., Pape, D., Teppke, M., "Zur Dunnschichtchromatographie Organischer Schwefelverbindungen von typ RSH and RSR′," *J. Chromatog.* (1965) **19,** 375.
(14) Orr, W. L., "Separation of Alkyl Sulfides by Liquid-Liquid Chromatography on Stationary Phases Containing Mercuric Acetate," *Anal. Chem.* (1966) **38,** 1558.
(15) Douglas, A. G., "Gas Chromatography," in "Organic Geochemistry," pp. 161–180, Springer-Verlag, New York, 1969.
(16) Sie, S. T., Rijnders, G. W. A., "High-Pressure Gas Chromatography and Chromatography with Supercritical Fluids. III. Fluid-Liquid Chromatography," *Separation Sci.* (1967) **2** (6), 729.
(17) Karayannis, N. M., Corwin, A. H., Baker, E. W., Klesper, E., Walter, J. A., "Apparatus and Materials for Hyperpressure Gas Chromatography of Nonvolatile Compounds," *Anal. Chem.* (1968) **40,** 1736.
(18) Burlingame, A. L., Schnoes, H. K., "Mass Spectrometry in Organic Geochemistry," in "Organic Geochemistry," pp. 89–160, Springer-Verlag, New York, 1969.
(19) Giddings, J. C., Mallik, K. L., "Theory of Gel Filtration (Permeation) Chromatography," *Anal. Chem.* (1966) **38,** 997.
(20) Rosscup, R. J., Pohlmann, H. P., "Molecular Size Distribution of Hydrocarbons and Metalloporphyrins in Residues as Determined by Gel Permeation Chromatography," "Abstracts of Papers," 153rd Meeting, ACS, April 1967, Petrol Q16.
(21) Rimington, C., Belcher, R. J., "Separation of Porphyrins on Sephadex Dextran Gels," *J. Chromatog.* (1967) **28,** 112.
(22) Blumer, M., Snyder, W. D., "Porphyrins of High Molecular Weight in a Triassic Oil Shale: Evidence by Gel Permeation Chromatography," *Chem. Geol.* (1967) **2,** 35.
(23) Hodgson, G. W., Holmes, M. A., Halpern, B., "Biogeochemistry of Molecular Complexes of Amino Acids with Chlorins and Porphyrins," *Geochim. Cosmochim. Acta* (1970) **34,** 1107.
(24) Shapiro, A. L., Vinuela, E., Maizel, J. V., "Molecular Weight Estimation of Polypeptide Chains by Electrophoresis in SOS–Polyacrylamide Gels," *Biochem. Biophys. Res. Comm.* (1967) **28,** 815.
(25) Stephens, R., Suetaak, W., Schatz, P. N., "Magneto-optical Rotatory Dispersion of Porphyrins and Phthalocyanines," *J. Chem. Phys.* (1966) **44,** 4592.
(26) Il'yina, A. A., Personov, R. I., "Line Emission Spectra of 1,12-Benzperylene and Its Identification in Some Natural Products," *Geokhimiya* **1962** (11), 963.
(27) Mauer, L. G., Parker, P. C., "A Study of the Geochemistry of Vitamin B_{12}," *Marine Sci.* (1968) **13,** 29.
(28) Kuykendall, W. E., Hislop, J. S., Traxler, R. N., "Application of Activation Analysis to Asphalts," *Am. Chem. Soc., Div. Petrol. Chem. Preprints* (April 1967), A-27.

(29) Loev, B., Snader, K. M., " 'Dry-Column' Chromatography. A Preparative Chromatographic Technique with the Resoluability of Thin-Layer Chromatography," *Chem. Ind. (London)* **1965,** 15.
(30) Mundschenk, H., "Extraktionschromatographische Trennung der Freien Porphyrine auf Tri-*n*-butylphosphat-saulen in pH-Gradienten," *J. Chromatog.* **1969,** 393.
(31) Kvenvolden, K. A., Peterson, E., Pollock, G. E., "Optical Configuration of Amino Acids in Precambrian Fig Tree Chert," *Nature* (1969) **221,** 141.
(32) Ray, B. R., Witherspoon, P. A., Grim, R. E., "A Study of the Colloidal Characters of Petroleum Using the Ultracentrifuge," *J. Chem. Phys.* (1957) **61,** 1296.
(33) Witherspoon, P. A., "Studies on Petroleum with the Ultracentrifuge," *Ill. State Geolg. Survey, Rept. Invest.* (1958) 206.
(34) Eldib, I. A., Dunning, H. N., Bolen, R. J., "Nature of Colloidal Materials in Petroleum," *J. Chem. Eng. Data* (1960) **5,** 550.
(35) Erdman, J. G., "The Molecular Complex Comprising Heavy Petroleum Fractions," in "Hydrocarbon Analysis," *A.S.T.M. S.T.P.* (1965) **389,** 259-300.
(36) Dickie, J. P., Yen, T. F., "Macrostructures of the Asphaltic Fractions by Various Instrumental Methods," *Anal. Chem.* (1967) **39,** 1847.
(37) Yen, T. F., Boucher, W., Dickie, J. P., Tynan, E. C., Vaughan, G. B. "Vanadium Complexes and Porphyrins in Asphaltenes," *J. Inst. Petrol.* (1969) **55,** 87.
(38) Bitz, M. C., Nagy, B., "Ozonolysis of Polymer-Type Material in Coal, Kerogen, and in the Orgueil Meteorite: A Preliminary Report," *Proc. Natl. Acad. Sci.* (1966) **56,** 1383.
(39) Douglas, A. G., Eglinton, G., Henderson, W., "Thermal Alteration of the Organic Matter in Sediments," in "Advances in Geochemistry," pp. 369–388, Pergamon, New York, 1966.
(40) Smith, G., Ellis, M. M., "Chromatographic Analysis of Gases from Soils and Vegetation Related to Geochemical Prospecting for Petroleum," *Bull. Am. Assoc. Petrol. Geol.* (1963) **47,** 1897.
(41) McCrossan, R. G., "An Evaluation of Surface Geochemical Prospecting for Petroleum, Caroline area, Alberta," Third International Geochemical Exploration Symposium, Toronto, Canada (1970).
(42) Deroo, G., Durand, B., Espitalie, J., Pelet, R., Tissot, B., "Possibilite d'Application des Modules Mathematiques de Formation du Petrole a la Propection dans les Bassins Sedimentaires," in "Advances in Organic Geochemistry," pp. 345–354, Pergamon, New York, 1969.
(43) Gerard, R. E., "Application of Data Processing Methods in Geochemical Prospecting for Petroleum," Third International Geochemical Exploration Symposium, Toronto, Canada (1970).
(44) McClusky, J. B., "Gas Geochemistry in the Seventies—Challenge and Concepts," Third International Geochemical Exploration Symposium, Toronto, Canada (1970).
(45) Hawkes, H. E., "Bibliography on Geochemical Prospecting," Third International Geochemical Exploration Symposium, Toronto, Canada (1970).
(46) Silverman, S. R., "Carbon Isotopic Evidence for the Role of Lipids in Petroleum Formation," *J. Am. Oil Chemists Soc.* (1967) **44,** 691.
(47) Thode, H. G., Monster, J., "Sulfur-Isotope Geochemistry of Petroleum, Evaporites, and Ancient Seas," in "Fluids in Subsurface Environments —A Symposium," *Am. Assoc. Petrol. Geol. Memoir* **4** (1965) 367.
(48) Hitchon, B., Friedman, I., "Geochemistry and Origin of Formation Waters in the Western Canada Sedimentary Basin—I. Stable Isotopes of Hydrogen and Oxygen," *Geochim. Cosmochim. Acta* (1969) **33,** 1321.

(49) Oakwood, T. S., "Report of Progress," in "Fundamental Research on Occurrence and Recovery of Petroleum, 1944–45," pp. 92–102, API, New York (cited by Davis) (1967).
(50) Kaneda, T., "Hydrocarbons in Spinach: Two Distinctive Carbon Ranges of Aliphatic Hydrocarbons," *Phytochem.* (1969) **8**, 2039.
(51) Hunt, J. M., Jamieson, G. W., "Oil and Organic Matter in Source Rocks," *Bull. Am. Assoc. Petrol. Geol.* (1956) **40**, 477.
(52) Hunt, J. M., "Distribution of Hydrocarbons in Sedimentary Rock," *Geochim. Cosmochim. Acta* (1961) **22**, 37.
(53) Peake, E., Hodgson, G. W., "Alkanes in Aqueous Systems. I. Exploratory Investigations on the Accommodation of C_{20}–C_{33} *n*-Alkanes in Distilled Water and Occurrence in Natural Water Systems," *J. Am. Oil Chemists Soc.* (1966) **43**, 215.
(54) Kriss, A. Ye., "Marine Microbiology (Deep Water)," *Izv. Akad. Nauk. S.S.S.R.* (1959).
(55) Postma, H., "Dissolved Organic Matter in the Oceans," in "Advances in Organic Geochemistry," pp. 47–57, Pergamon, New York, 1969.
(56) Jacobs, M. B., Ewing, M., "Suspended Particulate Matter: Concentration in the Major Oceans," *Science* (1969) **163**, 380.
(57) Manheim, F. T., Meade, R. M., Bond, G. C., "Suspended Matter in Surface Waters of the Atlantic Continental Margin from Cape Cod to the Florida Keys," *Science* (1970) **167**, 371.
(58) Kvenvolden, K. A., "Evidence for Transformations of Normal Fatty Acids in Sediments," in "Advances in Organic Geochemistry," pp. 335–366, Pergamon, New York, 1970.
(59) Kvenvolden, K. A., Hodgson, G. W., Peterson, E., Pollock, G. E., "Organic Geochemistry of the Swaziland System, South Africa," "Abstracts," G.S.A. Meeting, Mexico City (1968).
(60) Gregoire, C., "Experimental Diagenesis of the Nautilus Shell," in "Advances in Organic Geochemistry," pp. 429–441, Pergamon, New York, 1970.
(61) Meinschein, W. G., "Hydrocarbons—Saturated, Unsaturated, and Aromatic," in "Organic Geochemistry," pp. 330–356, Springer-Verlag, New York, 1969.
(62) Hills, I. R., Whitehead, E. V., "Triterpanes in Optically Active Petroleum Distillates," *Nature* (1966) **209**, 977–979.
(63) Holm-Hansen, O., "Algae: Amounts of DNA and Organic Carbon in Single Cells," *Science* (1969) **163**, 87.
(64) Vallentyne, J. R., "The Molecular Nature of Organic Matter in Lakes and Oceans, with Lesser Reference to Sewage and Terrestrial Soils," *J. Fish. Res. Bd., Canada* (1957) **14**, 33.
(65) Bottomley, W. B., "The Isolation from Peat of Certain Nucleic Acid Derivatives," *Proc. Roy. Soc.* (1917) **B90**, 39.
(66) Peterson, W. H., Fred, E. B., Domogalla, B. P., "The Occurrence of Amino Acids and Other Organic Nitrogen Compounds in Lake Water," *J. Biol. Chem.* (1925) **63**, 287.
(67) Rosenberg, E., "Purine and Pyrimidines in Sediments from the Experimental Mohole," *Science* (1964) **146**, 1680.
(68) Swain, F. M., Pakalns, G. V., Bratt, J. G., "Possible Taxonomic Interpretation of Some Paleozoic and Precambrian Carbohydrate Residues," in "Advances in Organic Geochemistry," pp. 469–491, Pergamon, New York, 1970.
(69) Hodgson, G. W., Peake, E., Baker, B. L., "The Origin of Petroleum Porphyrins: The Position of the Athabasca Oil Sands," in the "K. A. Clark Volume," pp. 75–100, Research Council of Alberta, Canada, 1963.

(70) Peake, E., Baker, B. L., Hodgson, G. W., "The Contribution of Amino Acids and Chlorins to the Beaufort Sea by the Mackenzie River System," Annual Meeting, Geological Society of America, 1970.
(71) Hodgson, G. W., Hitchon, B., Taguchi, K., Baker, B. L., Peake, E., "Geochemistry of Porphyrins, Chlorins and Polycyclic Aromatics in Soils, Sediments and Sedimentary Rocks," *Geochim. Cosmochim. Acta* (1968) **32,** 737.
(72) Baker, E. W., "Porphyrins" in "Organic Geochemistry," pp. 464–497, Springer-Verlag, New York, 1969.
(73) Murphy, M. T. J., "Analytical Methods," in "Organic Geochemistry," pp. 74–88, Springer-Verlag, New York, 1969.
(74) Kjaer, A., "The Distribution of Sulphur Compounds," in "Chemical Plant Taxonomy," pp. 453–473, Academic, New York, 1963.
(75) Adams, D. D., Richards, F. A., "Dissolved Organic Matter in an Anoxic Fjord, with Special Reference to the Presence of Mercaptans," *Deep Sea Res.* (1968) **15,** 471.
(76) Smith, P. V., "Studies on the Origin of Petroleum: Occurrence of Hydrocarbons in Recent Sediments," *Bull. Am. Assoc. Petrol. Geol.* (1954) **38,** 377.
(77) Davis, J. B., "Petroleum Microbiology," Elsevier, New York, 1967.
(78) Speers, G. C., Whitehead, E. V., "Crude Petroleum," in "Organic Geochemistry," pp. 638–675, Springer-Verlag, New York, 1969.
(79) Stevenson, F. J., Butler, J. H. A., "Chemistry of Humic Acids and Related Pigments," in "Organic Geochemistry," pp. 534–557, Springer-Verlag, New York, 1969.
(80) Bordovsky, O. K., "Accumulation and Transformation of Organic Substance in Marine Sediments," Parts 1–3, *Marine Geol.* (1965) **3,** 3.
(81) Bars, E. A., Fikham, A. N., "Determination of Naphthenic Acids in Natural Waters Containing Humic Acids," *Novosti Neft. Tekhn., Geol.* (1958) **7,** 36.
(82) Corbett, C. S., "In situ Origin of McMurray Oil of Northeastern Alberta and Its Relevance of General Problem of Origin of Oil," *Bull. Am. Assoc. Petrol. Geol.* (1955) **39,** 1601.
(83) Eglinton, G., Hamilton, R. J., "The Distribution of Alkanes," in "Chemical Plant Taxonomy," pp. 187–217, Academic, New York, 1963.
(84) Shorland, F. B., "The Distribution of Fatty Acids in Plant Lipids," in "Chemical Plant Taxonomy," pp. 253–303, Academic, New York, 1963.
(85) Kaneda, T., "Biosynthesis of Branched-Chain Fatty Acids. V. Microbial Steriospecific Syntheses of D-12-Methyltetradecanoic and D-14-Methylhexadecanoic Acids," *Biochim. Biophys. Acta* (1966) **125,** 43.
(86) Sorensen, N. A., "Chemical Taxonomy of Acetylenic Compounds," in "Chemical Plant Taxonomy," pp. 219–252, Academic, New York, 1963.
(87) Mahler, H. R., Cordes, E. H., "Basic Biological Chemistry," Harper and Rowe, New York, 1968.
(88) Lascelles, J., "Tetrapyrrole Biosynthesis and Its Regulation," Benjamin, New York, 1964.
(89) Calvin, M., Bassham, J. A., "The Photosynthesis of Carbon Compounds," Benjamin, New York, 1962.
(90) Davis, B. D., "On the Importance of Being Ionized," *Arch. Biochem. Biophys.* (1958) **78,** 497 (as cited in Ref. 89, p. 65).
(91) Hodgson, G. W., "Hydrocarbons," in "Encyclopedia of Earth Sciences," in press.
(92) Smith, H. M., Dunning, H. N., Rall, H. T., Ball, J. S., "Keys to the Mystery of Crude Oil," *Am. Petrol. Inst. Proc., Div. Refining,* New York (1959) **39,** 433.

(93) Rosenfeld, W. D., "Fatty Acid Transformations by Anaerobic Bacteria," *Arch. Biochem. Biophys.* (1948) **16,** 263.

(94) Mekhtieva, V. L., Kondrat'eva, G. F., Malkova, S. B., "The Transformation of Fats through the Action of Microorganisms; 1. The Decomposition of Fats during the Course of Denitrification," *Mikrobiol.* (1960) **29,** 85.

(95) Barker, H. A., "Studies on Methane Fermentation. VI. The Influence of Carbon Dioxide Concentration on the Rate of Carbon Dioxide Reduction by Molecular Hydrogen," *Proc. Natl. Acad. Sci.* (1943) **29,** 184.

(96) Farkas, L., Schneidnesser, B., "The Hydrogenation of Fumarate by Heavy Hydrogen in the Presence of *Bacillus coli,*" *J. Biol. Chem.* (1947) **167,** 807.

(97) Hoogerheide, J. C., Kocholaty, W., "Metabolism of the Strict Anaerobes (genus: Clostridium). 2. Reduction of Amino Acids with Gaseous Hydrogen by Suspension of *Clostridium sporogenes,*" *Biochem. J.* (1938) **32,** 949.

(98) ZoBell, C. E., "Role of Microorganisms in Petroleum Formation," *Am. Petrol. Inst., Res. Proj. 43A, Rept.* **39** (1952).

(99) Avigan, J., Blumer, M., "On the Origin of Pristane in Marine Organisms," *J. Lipid Res.* (1968) **9,** 350.

(100) Mair, B. J., "Terpenoids, Fatty Acids and Alcohols as Source Materials for Petroleum Hydrocarbons," *Geochim. Cosmochim. Acta* (1964) **28,** 1303.

(101) Winters, J. C., Williams, J. A., "Microbiological Alteration of Crude Oil in the Reservoir," *Am. Chem. Soc. Div. Petrol. Chem., Preprints* (Sept. 1969) **14,** (4), E22.

(102) Davis, J. B., Raymond, R. L., "Oxidation of Alkyl-Substituted Cyclic Hydrocarbons by a Nocardia during Growth on *n*-Alkanes," *Appl. Microbiol.* (1961) **9,** 383.

(103) Davis, J. B., Yarbrough, H. F., "Anaerobic Oxidation of Hydrocarbons by *Desulfovibrio desulfuricans,*" *Chem. Geol.* (1966) **1,** 137.

(104) Andreyevski, I. L., "The Action of Bacteria on an Oil Seam," *Piroda* **1958,** (10), 90 (*Biol. Abstr.* **14080**).

(105) Weiss, A., "Organic Derivatives of Clay Minerals, Zeolites and Related Minerals," in "Organic Geochemistry," pp. 737–781, New York, 1969.

(106) Weiss, A., Roloff, G., "Hamin-Montmorillonit und Seine Bedeutung fur die Festlegung der Oberen Temperaturgrenze bei der Bildung der Erdols," *Z. Naturforsch.* (1964) **19b,** 533.

(107) Kroepelin, H., "Racemisation of Amino Acids on Silicates," in "Advances in Organic Geochemistry," pp. 535–542, Pergamon, New York, 1968.

(108) Stevenson, F. J., Cheng, C-N., "Amino Acids in Sediments: Recovery by Acid Hydrolysis and Quantitative Estimation by a Colorimetric Procedure," *Geochim. Cosmochim. Acta* (1970) **34,** 77.

(109) Drost-Hansen, W., "Structure of Water Near Solid Interfaces," *Ind. Eng. Chem.* (1969) **61** (11), 10.

(110) Ahrens, L. H., "Ionization Potentials and Metal-Amino Acid Complex Formation in the Sedimentary Cycle," *Geochim. Cosmochim. Acta* (1966) **30,** 1111.

(111) Slowey, J. F., Jeffrey, L. M., Hood, D. W., "Evidence for Organic Complexed Copper in Sea Water," *Nature* (1967) **214,** 377.

(112) Erdman, J. G., Harju, P. H., "The Capacity of Petroleum Asphaltenes to Complex Heavy Metals," *J. Chem. Eng. Data* (1962) **8,** 252.

(113) Chowdhury, A. N., Bose, B. B., "Role of 'Humus Matter' in the Formation of Geochemical Anomalies," Third International Geochemical Exploration Symposium, Toronto, Canada, 1970.

(114) Levinson, A. A., "Hydrogeochemistry of Rivers in the Mackenzie Drainage Basin," Third International Geochemical Exploration Symposium, Toronto, Canada, 1970.
(115) Koshy, E., Desai, M. V. M., Ganguly, A. K., "Studies on Organometallic Interactions in the Marine Environment. Part 1. Interaction of Some Metallic Ions with Dissolved Organic Substances in Sea-water," *Curr. Sci. (India)* (1969) **38,** 555.
(116) *Ibid.,* pp. 582–586.
(117) Mueller, G., "The Theory of Genesis of Oil through Hydrothermal Alteration of Coal-Type Substances within Certain Lower Carboniferous Strata of the British Isles," International Geological Congress, Algiers, Paper No. 12, 19th Session, 1954.
(118) Jensen, M. L., "Stable Isotopes in Geochemical Prospecting," Third International Geochemical Exploration Symposium, Toronto, Canada, 1970.
(119) Fabuss, B. M., Duncan, D. A., Smith, J. O., Satterfield, C. N., "Effect of Organosulfur Compounds on the Rate of Thermal Decomposition of Selected Saturated Compounds," *Ind. Eng. Chem., Process Design Develop.* (1965) **4,** (1), 117.
(120) Douglas, A. G., Mair, B. J., "Sulfur: Role in Genesis of Petroleum," *Science* (1965) **147,** 499.
(121) Tucker, J. R., Schweyer, H. E., "Distribution and Reactions of Sulfur in Asphalt during Air Blowing and Sulfurizing Processes," *Ind. Eng. Chem. Prod. Res. Develop.* (1965) **4** (1), 51.
(122) Toland, W. G., "Oxidation of Alkylbenzenes with Sulfur and Water," *J. Org. Chem.* (1961) **26,** 2929.
(123) Peake, E., Hodgson, G. W., "Laboratory Studies of the Disaccommodation of *n*-Alkanes from Simulated Formation Waters," "Abstracts of Papers," Meeting, ACS, 1968.
(124) Billings, G. K., Hitchon, B., Shaw, D. R., "Geochemistry and Origin of Formation Waters in the Western Canada Sedimentary Basin. 2. Alkali Metals," *Chem. Geol.* (1969) **4,** 211.
(125) Whitehead, E. V., "Molecular Evidence for the Biogenesis of Petroleum and Natural Gas," International Symposium on Hydrogeochemistry and Biogeochemistry, Tokyo, 1970.
(126) Yen, T. F., Silverman, S. R., "Geochemical Significance of the Changes in the Chemical Structures of the Complex Heterocyclic Compounds of Petroleums," *Am. Chem. Soc. Div. Petrol. Chem., Preprints* (Sept. 1970) E32.
(127) Bendoraitis, J. G., Brown, B. L., Hapner, L. S., "Isoprenoid Hydrocarbons in Petroleum. Isolation of 2, 6, 10, 14-Tetramethylpentadecane by High-Temperature Gas-Liquid Chromatography," *Anal. Chem.* (1962) **34,** 49.
(128) Hodgson, G. W., Baker, B. L., Peake, E., "The Role of Porphyrins in the Geochemistry of Petroleum," *Proc. World Petrol. Congr., 7th,* Mexico City (1967) **9,** 117.
(129) Andreev, P. F., Bogomolov, Dobryanskii, A. F., Kartsev, A. A., "Transformation of Petroleum in Nature," Pergamon, New York, 1968 (transl. by R. B. Gaul and B. C. Metzner.
(130) Cheshire, M. V., Cranwell, P. A., Falshaw, C. P., Floyd, A. J., Haworth, R. D., "Humic Acid—II. Structure of Humic Acids," *Tetrahedron* (1967) **23,** 1669.
(131) Jurg, J. W., Eisma, E., "Petroleum Hydrocarbons: Generation from Fatty Acids," *Science* (1964) **144,** 1451.
(132) Eisma, E., Jurg, J. W., "Fundamental Aspects of the Generation of Petroleum," in "Organic Geochemistry," pp. 676–698, Springer-Verlag, New York, 1969.

(133) Jurg, J. W., Eisma, E., "The Mechanism of the Generation of Petroleum Hydrocarbons from a Fatty Acid," (abstract) in "Advances in Geochemistry," Pergamon, New York, 1970.
(134) Erdman, J. G., "Some Chemical Aspects of Petroleum Genesis Related to the Problem of Source-Bed Recognition," *Geochim. Cosmochim. Acta* (1961) **22,** 16.
(135) Silverman, S. R., "Investigations of Petroleum Origin and Evolution Mechanisms by Carbon Isotope Studies," in "Isotopic and Cosmic Chemistry," North Holland, Amsterdam, 1964.
(136) Smith, H. M., Ball, J. S., "The Effect of Time and Depth of Burial on the Naphtha and Gas Oil Content of Crude Oils," *Am. Chem. Soc. Div. Petrol. Chem., Preprints* (Sept. 1969) E5.
(137) Oudin, J. L., Califet-Debyser, Y., "Influence of Depth and Temperature on the Structure and Distribution of Alkanes and Aromatic Hydrocarbons from Rock Extracts," *Am. Chem. Soc. Div. Petrol. Chem., Preprints* (Sept. 1969) E16.
(138) Dayhoff, M. O., Lippincott, E. R., Eck, R. J., Nagarajan, G., "Thermodynamic Equilibrium in Prebiological Atmospheres of C, H, O, N, P, S, and Cl," *NASA SP*-**3040** (1967).
(139) Hodgson, G. W., Baker, B. L., "Vanadium, Nickel and Porphyrins in Thermal Geochemistry of Petroleum," *Bull. Amer. Assoc. Petrol. Geol.* (1957) **41,** 2413.
(140) Rosscup, R. J., Bowman, "Thermal Stabilities of Vanadium and Nickel Petroporphyrins," "Abstracts of Papers," 153rd Meeting, ACS, 1967.
(141) Vallentyne, J. R., "Biogeochemistry of Organic Matter—II. Thermal Reaction Kinetics and Transformation Products of Amino Compounds," *Geochim. Cosmochim. Acta* (1964) **28,** 157.
(142) Vallentyne, J. R., "Pyrolysis of Proline, Leucine, Arginine and Lysine in Aqueous Solution," *Geochim. Cosmochim. Acta* (1968) **32,** 1353.
(143) Schopf, J. W., Kvenvolden, K. A., Barghoorn, E., "Amino Acids in Precambrian Sediments: An assay," *Proc. Natl. Acad. Sci.* (1968) **59,** 639.
(144) Swain, F. M., Bratt, J. M., Kirkwood, S., "Carbohydrates from Precambrian and Cambrian Rocks and Fossils," *Geol. Soc. Am., Ann. Meeting* (1969).
(145) Nagy, B., "Porosity and Permeability of the Early Precambrian Onverwacht Chert: Origin of the Hydrocarbon Content," *Geochim. Cosmochim Acta* (1970) **34,** 525.
(146) Evans, C. R., "Regional Facies of Organic Metamorphism," Third International Geochemical Exploration Symposium, Toronto, Canada, 1970.
(147) Weaver, C. E., "The Significance of Clay Minerals in Sediments," in "Fundamental Aspects of Petroleum Geochemistry," pp. 37–75, Elsevier, New York, 1967.
(148) Mitterer, R. M., "Amino Acid Composition of Organic Matrix in Calcareous Oolites," *Science* (1968) **162,** 1498.
(149) Baker, B. L., Hodgson, G. W., "Rate of Formation of the Nickel Complex of *Pheophytin a.,*" *J. Phys. Chem.* (1961) **65,** 1078.
(150) Hodgson, G. W., Hitchon, B., Elofson, R. M., Baker, B. L., Peake, E., "Petroleum Pigments from Recent Freshwater Sediments," *Geochim. Cosmochim. Acta* (1960) **19,** 272.
(151) Swain, F. M., "Fossil Carbohydrates," in "Organic Geochemistry," pp. 374–400, Springer-Verlag, New York, 1969.
(152) Degussa Co., "Facelift for Silica," *Ind. Eng. Chem.* (1968) **60** (12), 6.
(153) Hodgson, G. W., Hitchon, B., "Primary Degradation of Chlorophyll under Simulated Petroleum Source Rock Sedimentation Conditions," *Bull. Am. Assoc. Petrol. Geol.* (1968) **43,** 2481.

(154) Dickey, P. A., University of Tulsa, private communication (1968).
(155) McAuliffe, C., "Solubility in Water of C_1–C_9 Hydrocarbons," *Nature* (1963) **200**, 1092.
(156) McAuliffe, C., "Solubility in Water of Paraffin, Cycloparaffin, Olefin, Acetylene, Cycloolefin, and Aromatic Hydrocarbons," "Abstracts of Papers," 144th Meeting, ACS (1964).
(157) McAuliffe, C., "Gas Chromatography Determination of Dissolved Hydrocarbons in Various Aqueous Media by Repeated Equilibration of Sample with Gas," CIC-ACS Joint Conference, Toronto, Canada, May 1970, Anal. 5.
(158) Franks, F., "Solute-Water Interactions and the Solubility Behavior of Long Chain Paraffin Hydrocarbons," *Nature* (1966) **210**, 87.
(159) Baker, E. G., "Network Topology and Cohesive Forces in Polyethylene," International Symposium on Macromolecular Chemistry, Toronto, Canada, 1968.
(160) Baker, E. G., "Origin and Migration of Oil," *Science* (1959) **129**, 871.
(161) Baker, E. G., "A Hypothesis Concerning the Accumulation of Sediment Hydrocarbons To Form Crude Oil," *Geochim. Cosmochim. Acta* (1960) **19**, 309.
(162) Baker, E. G., "Distribution of Hydrocarbons in Petroleum," *Bull. Am. Assoc. Petrol. Geol.* (1962) **46**, 76.
(163) Baker, E. G., "A Geochemical Evaluation of Petroleum Migration and Accumulation," in "Fundamental Aspects of Petroleum Geochemistry," pp. 299–329, Elsevier, New York, 1967.
(164) Johansen, R. T., Dunning, H. N., "Relative Wetting Tendencies of Crude Oils by Capillarimetric Method," *U.S. Bur. Mines Rept. Invest.* **5752** (1961).
(165) Moore, J. W., Dunning, H. N., "Interfacial Activities and Porphyrin Contents of Oil-Shale Extracts," *Ind. Eng. Chem.* (1955) **47**, 1440.
(166) Frank, S. G., Zograti, G., "Solubilization of Water by Dialkyl Sodium Sulfosuccinates in Hydrocarbon Solutions," *J. Coll. Interface Sci.* (1969) **29** (1), 27.
(167) Davis, J. B., "Distribution of Naphthenic Acids in an Oil-Bearing Aquifer," *Chem. Geol.* (1969) **5**, 89.
(168) Pedersen, C. J., "Dissolving Salt in Benzene," *J. Am. Chem. Soc.* (1970) **92**, 391.
(169) Sokolov, V. A., Zhuse, T. P., Vassoyevich, N. B., Antonov, P. L., Grigoriyev, G. G., Kozlov, V. P., "Migration Processes of Gas and Oil, Their Intensity and Directionality," *Proc. World Petrol. Congr., 6th,* Sect. 1, Paper 47, 1963.
(170) Meinschein, W. G., Sternberg, A. M., Klusman, D. W., "Origins of Natural Gas and Petroleum," *Nature* (1968) **220**, 1185.
(171) O'Sullivan, T. D., Smith, N. O., Nagy, B., "Solubility of Natural Gases in Aqueous Salt Solutions—III Nitrogen in Aqueous NaCl at High Pressures," *Geochim. Cosmochim. Acta* (1966) **30**, 617.
(172) Hedberg, H. D., "Significance of High-Wax Oils with Respect to Genesis of Petroleum," *Bull. Am. Assoc. Petrol. Geol.* (1968) **52**, 736.
(173) Hodgson, G. W., Hitchon, B., Taguchi, K., "The Water and Hydrocarbon Cycles in the Formation of Oil Accumulations," in "Recent Researches in the Fields of Hydrosphere, Atmosphere and Nuclear Geochemistry," pp. 217–242, Maruzen, Tokyo, 1964.
(174) Hodgson, G. W., Hitchon, B., "Research Trends in Petroleum Genesis," Commonwealth Mining and Metallurgical Congress, 8th, Australia, March 1965, Paper No. 34.
(175) Hitchon, B., "Rock Volume and Pore Volume Data for Plains Region of Western Canada Sedimentary Basin between Latitudes 49° and 60°N," *Bull. Am. Assoc. Petrol. Geol.* (1968) **52**, 2318.

(176) Hitchon, B., "Fluid Flow in Western Canada Sedimentary Basin. 1. Effect of Topography," *Water Resources Res.* (1969) **5,** 186.
(177) Hitchon, B., "Fluid Flow in Western Canada Sedimentary Basin. 2. Effect of Geology," *Water Resources Res.* (1969) **5,** 460.
(178) Brezgunov, V. S., Vlasova, L. S., Soyfer, V. N., "Isotopic Composition of Hydrogen as a Clue to the Origin of Ground Waters and Petroleum," *Geokhimiya* **1968,** 86.
(179) Silverman, S. R., "The Migration and Segregation of Oil and Gas," Geology of Fluids Symposium, Midland, Tex., Jan. 1964.
(180) Dott, R. H., Reynolds, M. J., "Sourcebook for Petroleum Geology," *Am. Assoc. Petrol. Geol.* (1969) *Memoir* **5.**
(181) Bonham, L. C., "Geochemical Investigation of Crude Oils," *Bull. Am. Assoc. Petrol. Geol.* (1956) **40,** 897.
(182) Nagy, B., Wourms, J. P., "Experimental Study of Chromatographic-Type Accumulation of Organic Compounds in Sediments. An Introductory Statement," *Bull. Geol. Soc. Am.* (1959) **70,** 655.
(183) Hodgson, G. W., Baker, B. L., "Geochemical Aspects of Petroleum Migration in Pembina, Redwater, Joffre, and Lloydminster Oil Fields of Alberta and Saskatchewan, Canada," *Bull. Am. Assoc. Petrol. Geol.* (1959) **43,** 311.
(184) Philipp, W., Brong, H. J., Fuchthauer, H., Haddenhorst, H. G., Jankowsky, W., "The History of Migration in the Gifhorn Trough (N.W.-Germany)," *Proc. World Petrol. Congr., 6th,* Section 1, Paper 19, Panel Discussion 2 (1963).
(185) Herzig, J. P., Leclerc, D. M., LeGoff, P., "Flow of Suspensions through Porous Media—Application to Deep Filtration," *Ind. Eng. Chem.* (1970) **62** (5), 8.
(186) Rice, P. A., Fontugne, D. J., Latini, R. G., Barduhn, "Anisotropic Permeability in Porous Media," *Ind. Eng. Chem.* (1970) **64** (6), 23.
(187) Morrow, N. R., "Physics and Thermodynamics of Capillary Action in Porous Media," *Ind. Eng. Chem.* (1970) **62** (6), 32.
(188) Carrier, G. F., "The Mixing of Ground Water and Sea Water in Permeable Subsoils," *J. Fluid Mech.* (1958) **4,** 479.
(189) Roberts, W. H., "Hydrodynamic Analysis in Petroleum Exploration," "Encyclopedia del Petrolio e die Gas Naturali," in press.
(190) Nigrini, A., "Investigations into the Transport and Deposition of Copper, Lead and Zinc in the Surficial Environment," Third International Geochemical Exploration Symposium, Toronto, Canada, 1970.
(191) Jeffrey, D. A., Zarrella, W. M., "Geochemical Prospecting at Sea," *Am. Assoc. Petrol. Geol. Meetg.,* Calgary, Canada, June 1970.
(192) Bonoli, L., Witherspoon, P. A., "Diffusion of Paraffin, Cycloparaffin and Aromatic Hydrocarbons in Water and Some Effects of Salt Concentration," in "Advances in Organic Geochemistry," Pergamon, New York, 1969.
(193) Sokolov, V. A., "The Theoretical Foundations and Development Tendencies of Geochemical Prospecting for Petroleum and Natural Gas," Third International Geochemical Exploration Symposium, Toronto, Canada, 1970.
(194) Sokolov, V. A., Vainbaum, S. J., Geodekjan, A. A., Grigoriev, G. G., Krems, A. J., Stroganov, V. A., Zorkin, L. M., Zeidelson, M. I., "The New Methods of Gas Surveys; Gas Investigations of Wells and Some Practical Results," Third International Geochemical Exploration Symposium, Toronto, Canada, 1970.

RECEIVED June 26, 1970.

2

Origin of Oil: Geological and Geochemical Constraints

BRIAN HITCHON

Senior Research Officer, Research Council of Alberta,
Edmonton, Alberta, Canada

The geological and geochemical constraints on the origin of oil are broad and minimal—namely, generation of a mobile phase from organic matter in the sedimentary environment through low temperature reactions and emplacement within that sedimentary environment (or subjacent igneous or metamorphic rocks) in situations where the geochemical characteristics of the aqueous carrier fluid cause unloading of the mobile organic phase to form oilfields. The vicissitudes of structural and stratigraphic variations within the sedimentary environment and the changes of permeability and porosity with time are all factors complexing an otherwise relatively simple situation—they are not constraints, as the number of dry structural and stratigraphic "traps" testifies. The origin, maturation, migration, and accumulation of oil are discussed within this framework and supported by 303 references.

There is usually a delay of several years between the publication of new concepts in a research paper and their incorporation in standard textbooks. The specialists in petroleum geology and petroleum geochemistry are aware of the recent research which expands on the thoughts expressed in standard textbooks (*1, 2, 3, 4, 5, 6*). They are also familiar with the history of geological thought as applied to petroleum (*7*). This review paper, and its companion review paper (*8*), are directed, therefore, at the nonspecialist. They will attempt to define the geological and chemical constraints, respectively, within which we believe crude oil originates.

It has become almost a convention of papers reviewing the origin of oil to discuss this subject under the broad headings, origin, migration, and accumulation. This author has been no exception (*9*). However, because many of the chemical aspects are covered in the companion paper (*8*), this review is restricted generally to the geological and geochemical aspects of the origin of oil and discusses:

(a) The fate of organic matter in the geological environment, with particular emphasis on petroleum.

(b) The maturation of organic matter and petroleum with time and temperature.

(c) The migration and accumulation of petroleum in an aqueous environment and under the fluid pressure, thermal, and chemical gradients which exist in sedimentary basins.

(d) Organic-inorganic reactions in the geological environment and the possible relations between oil and ore deposits in sedimentary basins.

The complex nature of this subject suggests the need for a lengthy review. This has been avoided by the liberal use of references to which the reader is referred for more comprehensive coverage of the geological and geochemical constraints on the origin of oil.

Organic Matter in the Geological Environment

The literature on organic geochemistry has increased considerably during the past decade. In addition to the papers cited individually there have been seven special volumes of papers (*10–16*), two important translations from the Russian literature (*17, 18*), and a book dealing with petroleum microbiology (*19*). Many review papers have dealt with the organic geochemical aspects of petroleum (*9, 20–39*).

An organic origin for petroleum is assumed because of:

(a) Geological association.

(b) The presence of biological markers, some of which are optically active.

(c) Carbon isotope studies.

(d) Similarity of compounds of petroleum and living organisms

The majority of oil and gas occurrences and their presumed source rocks are associated with sedimentary strata deposited during periods when there was abundant life. The absence of evidence of life in all igneous and most Early Precambrian rocks coincides with a paucity of petroleum occurrences. The relation of abundance of life and occurrence of petroleum throughout the Phanerozoic is shown in Figure 1. The relative abundance of three major groups of organisms is represented by curves showing the number of extant taxa (left curve) and total taxa present (right curve) over geologic time and is compared with the number of taxa for all organisms (*40*) and the relative amounts of petroleum discovered in major fields in the free world (*41*). The curves for all organisms suggests that the Tertiary would be particularly favorable for

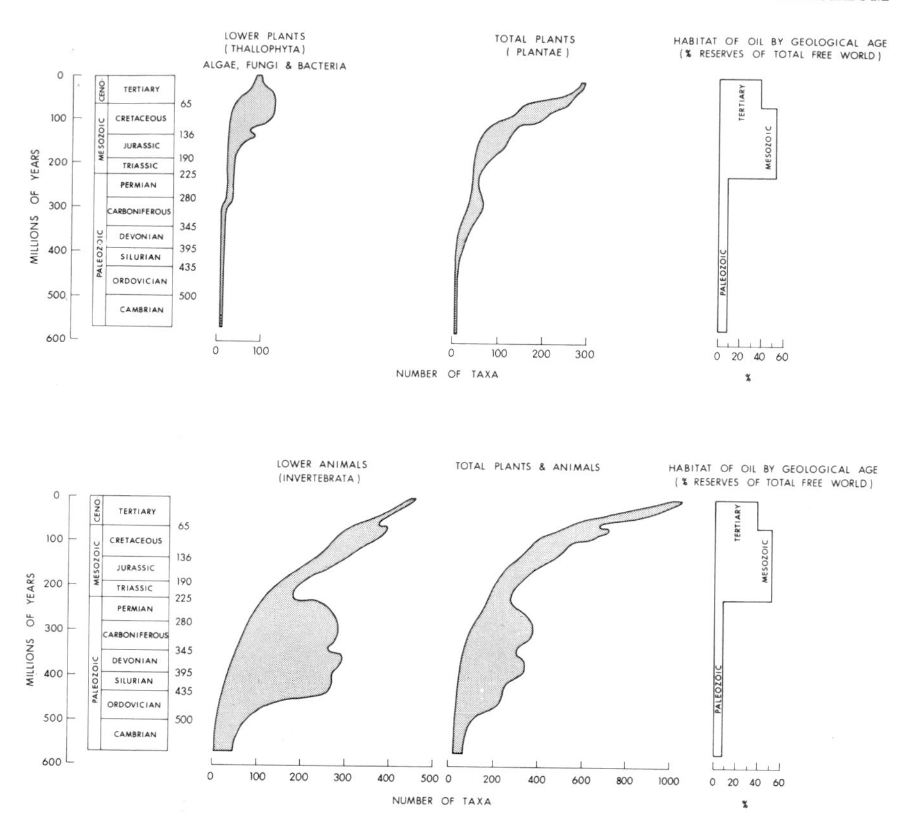

Figure 1. Relation of the variation in number of taxa of various groups of organisms over geologic time (40) to the amount of petroleum in the major oil fields of the free world (41). The difference between the curve for total taxa (right) and curve for extant taxa (left) indicates the number of taxa that became extinct for that period of geologic time.

petroleum, and Weeks (*42*) has noted that the "upper half of the Tertiary has to date accounted for upward of 35 per cent of the oil found in the world." In general, the Mesozoic should also contain significant quantities of petroleum, with the Paleozoic the least important unit. The shape of the curves for the three major groups of organisms strongly suggests that it is the abundance of plants, and particularly the lower plants, which correlates best with the curve of petroleum occurrence, and it is particularly in the lower plants that we find a large biomass of potential hydrocarbon-producing material. We note later the convincing geochemical evidence now available to support this statement. The majority of the strata in which petroleum has been found were deposited under marine conditions although sediments of super saline, brackish, or even fresh-water environments may yield significant amounts of petroleum. Indeed, the interrelationships of the environmental spectrum from marine petroleum, through brackish conditions, to nonmarine coal have not yet been completely elucidated. However, the more favorable conditions would seem to be a reducing marine environment with an absence of bacteria which would completely destroy organic matter, or at least active and rapid deposition of fine-grained sediments—intertongued with potential reservoir rocks such as sands and carbonate reefs. Odum and Hoskin (*43*) have suggested that an additional factor may be the overproduction of food by photosynthesis, relative to the food consumed in the various food chains, and its consequent accumulation in the bottom sediments. However, the conditions that are operative are such that there are important losses of oxygen and nitrogen and gains of carbon and hydrogen as the organic material is transformed into petroleum (*33*). Details of transformations of individual groups of compounds are discussed in the next few paragraphs.

Such conditions are particularly well exemplified in the hinge-belts of major downwarps, as for example, the east-west Tethyan belt of Eurasia (North Africa-Caucasus-Caspian Sea-Middle East-Indonesia) or the eastern Circum-Pacific belt (Western Canada-Rockies-California-Mid-Continent-Gulf of Mexico-Central America-Venezuela-Columbia-Peru-Argentina). The minerals with which the organic matter is intimately associated throughout its history thus include calcite, dolomite, halite, gypsum (anhydrite), quartz, and clay minerals. The processes occurring at the mineral–organic interface are barely understood and would form the subject of many valuable research projects. Since solution and deposition of all these minerals may also take place in the underground environment, their history and relation to the movement of formation water—and the contained petroleum—are vital topics of interest if we are to understand fully the origin of petroleum. These and other geological aspects of the origin of petroleum are considered more completely in the textbooks cited in the introduction and in excellent summaries provided by a number of authors (*25, 26, 44, 45*). We shall therefore accept the geological association of oil and gas occurrences, sedimentary strata, and life processes.

Confirmation of the fundamental relation of life processes and petroleum is obtained by studying biological markers. Eglinton (*46*) has defined a biological marker as "a compound, the structure of which can be interpreted in terms of a previous biological origin." He has pointed out that to be useful the compound must have sufficient stability to survive long periods of time and sufficient complexity of structure (positional, and relative and absolute configuration) to render it distinctive.

Seldom is it a compound originally present in the organisms but more usually a related structure with the right carbon skeleton, absolute stereochemistry, and magnitude of rotation of polarized light (if it is an optically active compound)—*i.e.*, it must be the correct enantiomer. When a series of similar compounds, such as the acyclic isoprenoids, are all biological markers, further confirmation of their organic origin may be obtained by comparing the carbon number dominance in the living organisms with those compounds present in the rocks or petroleum being examined. Finally, the isotopic compositions of the original compounds must be maintained in those portions of the molecules which have remained intact and have not been degraded or rearranged. Examples of series of compounds which are useful biological markers are the acyclic isoprenoid alkanes and fatty acids, steranes and triterpanes, and the porphyrins. Figure 2 shows the carbon skeletons of some of these biological markers, and Table I summarizes our knowledge of their distribution in the geological environment. Several papers on the broad topic of biological markers have been published recently (*47–52*).

Carbon isotope studies of individual compounds present in organic matter in the geological environment have been reported only for the higher *n*-paraffins in three crude oils (*53*). Tentative interpretations from this study indicate that minima in the $^{13}C/^{12}C$ ratio curves plotted *vs.* carbon number, correspond to those *n*-paraffins which contain the highest concentration of biologically produced carbon skeletons. Conversely, maxima in the $^{13}C/^{12}C$ ratio curves indicate *n*-paraffins mainly of secondary origin. Broad groups of compounds in petroleum have been examined for their carbon isotopic composition (*54, 55*) and, together with studies of whole crude oils (*54–57*) have confirmed a biological origin for petroleum.

The evidence presented in the three preceeding paragraphs confirms a biological origin for petroleum. Since petroleum varies in composition, it is interesting to speculate if at least part of the variation results from differences in the composition of the organic matter from which it was derived. Research during the past decade suggests that variations could result from the effects of biochemical evolution, chemical taxonomy, the proportions of components from different groups of organisms, and the degree of maturation prior to petroleum generation.

Biochemical evolution refers to changes over geologic time of the fundamental composition of organic components—*e.g.*, the sequence of amino acids in protein molecules. The best documented example of biochemical evolution is that of the respiratory pigment haemoglobin, and the relation of its evolution to the fossil record has been summarized recently (*58*). Many of the monographs on comparative biochemistry have discussed biochemical evolution (*59, 60, 61*), and the reader is referred to them for possible conclusions of direct application to the origin of organic matter and petroleum. However, the present evidence suggests that while biochemical evolution has taken place, its effect on the composition of either the organic matter in sedimentary rocks or the derived petroleum is minimal.

When comparative biochemistry coincides with comparative biology in a group of animals or plants, they are said to exhibit chemical taxonomy. Probably the best known example of chemical taxonomy, in its broadest sense, that is relevant to the origin of petroleum, is the limitation of chlorophyll to plants and haemoglobin to the higher animals—both pigments are porphyrins (*62*). The presence of chlorophyll derivatives in the geological environment (*63*) and particularly in petroleum (*64*), indicates a predominantly plant origin for organic matter and petroleum. Triterpenes are extremely rare in the animal kingdom but

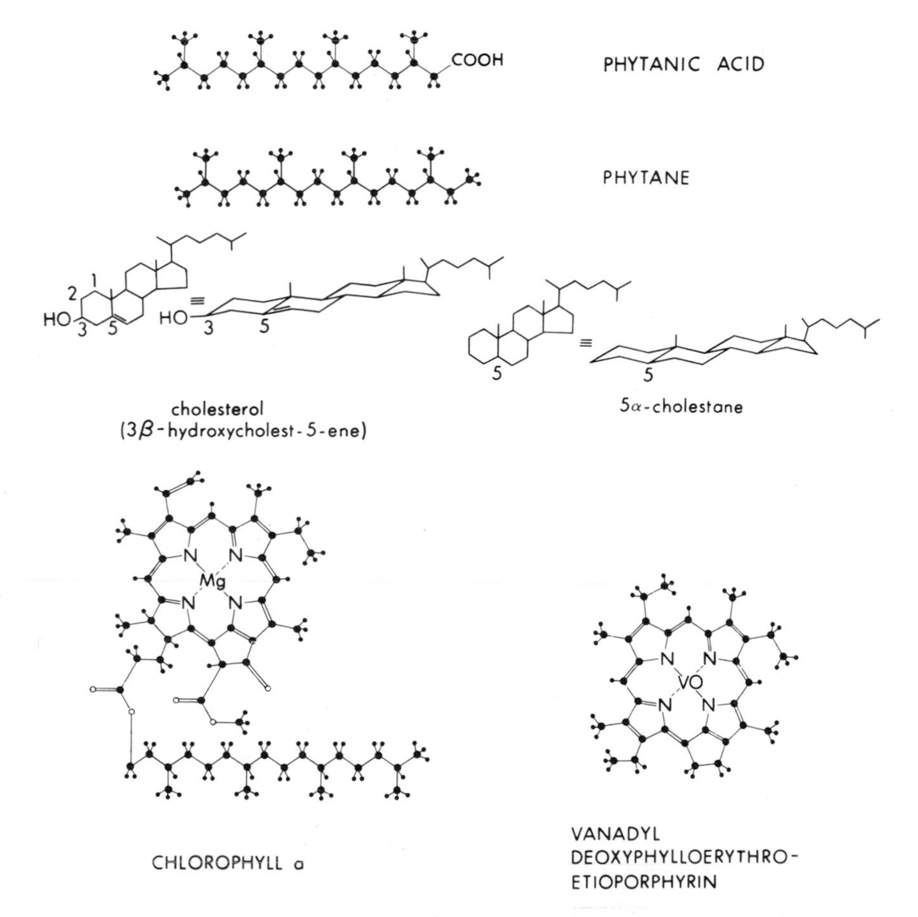

Figure 2. Carbon skeletons of selected biological markers

occur widely in the plant kingdom—a distribution possibly related to the absence of chlorophyll, and hence photosynthesis, in animals (*65*). Triterpanes, derivatives of triterpenes, are found in petroleum (*see* Table I), indicating a plant origin. The demonstration of chemical taxonomy in the geological environment has been a relatively recent accomplishment, and already some interesting geological implications are apparent. Studies of amino acids in Recent (*66, 67*) and fossil shells (*48, 68, 69, 70*) indicate that the amino acid composition is controlled by both phylogenic and environmental factors. A similar situation may prevail for other plant and animal fossils (*71*). Carbohydrate residues of fossil plants may also be of taxonomic significance (*72*). Work on living organisms suggests that alkanes (*73, 74, 75*) and fatty acids (*76*) in the lower plants such as bacteria, fungi, and algae are of phylogenic significance, and these biological groups are believed to be very important in the origin and diagenesis of the organic matter from which petroleum is derived. Thus the significance of chemotaxonomic variations in composition of the original living organisms to the organic matter incorporated into the geological environment and the subsequent effects on the petroleum derived from it are only beginning to be known.

From the discussion so far we can conclude that the organic matter —and petroleum—in the geological environment is derived mainly from plants. This is what we might have anticipated by knowing the close association of the majority of petroleum occurrences with marine sedimentary rocks and the dominant part played by phytoplankton in the biomass of the present day oceans. However, it is of some practical geological importance to know if terrigenous plant detritus has contributed significantly to the organic matter incorporated into the marine environment. After all, much of the inorganic detritus is land derived, and in some geological environments we may expect a significant portion of land-derived organic detritus. Using mainly geological evidence, Hedberg (*77*) suggested the association of high wax crude oils with a contribution of terrigenous organic matter. Subsequent geological studies of high wax crude oils (*78, 79*) and of isoprenoids (*80*) and triterpanes (*65, 81*) in petroleum lend support to the suggestion (*46*) that these compounds may be useful as environmental markers. Welte (*53*) has indicated that carbon isotope studies of individual compounds in crude oils can be used to advantage in deciphering the relative contributions of terrigenous and marine plants to a crude oil.

Summarizing the origin of organic matter in the geological environment, with particular reference to petroleum, we should note the convincing evidence based on geological, chemical, and isotopic studies for a biological origin, mainly from marine plants, with some contribution from terrigenous plants. Biochemical evolution and chemical taxonomy have both contributed to the complex composition of organic matter reaching the geological environment. The organic matter is incorporated into sediments by burial, and although much of it is lost by bacterial action, a significant portion is preserved, with preservation usually assisted by rapid sedimentation and the reducing conditions present at depth. Increased depth of burial is accompanied by increased maturation which affects both the immobile kerogenic material and the mobile petroleum—which has been generated from the total inorganic matter by maturation. Although the author does not subscribe to an inorganic origin for oil, it would be remiss in a review to omit mention that arguments have been brought forward to support such an origin. Recent papers on this aspect of the origin of oil include those by Robinson (*82*) and Rudakov (*83, 84*). We now discuss the geological implications of the maturation processes for it is only after extensive maturation that mobile petroleum is formed.

Maturation Process

The presence of biological markers in sedimentary rocks and petroleum does not imply that the organic matter in the rocks and the petroleum derived from it are merely accumulations of unaltered biological products. Erdman (*85*) has stated that living organisms do not synthesize, or at least do not commonly synthesize, certain compounds or families of compounds which are abundant in petroleum and the organic matter in sedimentary rocks. These include the low molecular weight paraffins up to the hexanes (*86*), the alkyl substituted benzene series below cymene, the lower members of the naphthalenic series including the parent compounds benzene and naphthalene, and the polynuclear aromatic hydrocarbons, anthracene, phenanthrene, and perylene. In addition, there is the enormous quantity of poorly defined asphaltic material.

This observation suggests that at least these compounds are mainly generated within the sedimentary rocks, although low molecular weight depth. Although radioactivity and the catalytic action of clays (*92–96*) modern shells (*90*). Rogers and Koons (*91*) have shown that oil forms by incorporation of heavy (C_{15+}) hydrocarbons from pre-existing organic matter and by generation of both light (C_4–C_7) and heavy (C_{15+}) hydrocarbons from bulk organic matter. The light hydrocarbons are absent in Recent sediments and are formed only over finite intervals of time and depth. Although radioactivity and the catalytic action of clays (*92, 96*) may play a part in the maturation of petroleum and the generation of new compounds, it is generally recognized that time and temperature are the main factors controlling the low temperature chemical reactions which are characteristic of the maturation of organic matter and petroleum.

Table I. Biological Markers

Biological Marker	*Living Matter*	*Recent Sediments*
Acyclic isoprenoid alkanes	Present in the marine environment but often unsaturated (*250, 251, 252*, (*253*)	Saturated forms reported from recent marine sediments (*254*)
Acyclic isoprenoid fatty acids	Abundant and widely distributed generally unsaturated (*46*)	First reported in recent marine sediments in 1967 (*275*); have since been found in several marine basins (*140, 141*). Cooper and Blumer (*140*) suggested that the isoprenoid acids may provide markers for the contribution of animals lipids to the organic matter of recent sediments
Steranes and triterpanes	Sterols (*280*) occur widely in the animal kingdom; phytosterols found in most groups of plants. Triterpenes are rare in animals but occur in many groups of plants. Pentacyclic triterpenes are in abundance and considerable variety in the higher land plants (*65*).	Sterols were first found in recent marine sediments in 1964 (*281*); since been reported by other workers (*282, 283*). No affirmative findings are available on triterpenes in recent sediments. Lipids form up to 20% of soil humus (*284*).
Porphyrins	Chlorins are the respiratory pigments in plants, and haemoglobin the respiratory pigment in the higher animals (*62, 64*).	Chlorins (both uncomplexed and metal complexed) have been known from recent freshwater and marine sediments for many years, but metal-complexed petroleum-type porphyrins have been found in only one freshwater sediment (*292*).

in the Geological Environment

Ancient Sedimentary Rocks	*Petroleum*
Widespread in a variety of rocks ranging in age from Late Pleistocene to Early Precambrian (*63, 255-264*), and including the Green River, Colo. and Scottish oil shales (*265-267*).	First reported from petroleum in 1963 (*268, 269*), acyclic isoprenoid alkanes have now been identified in crude oils from three continents (*270-273*), and in one crude oil all 12 isoprenoid hydrocarbons from C_{14} to C_{25} as well as head-to-tail conformations, have been found (*274*).
Only the Green River oil shale and Scottish torbanite have so far yielded acyclic isoprenoid fatty acids (*129, 143, 276-278*).	Have only been isolated from one California crude oil (*279*).
Sterols have been identified from a 2000-year old marine sediment (*283*) and both steranes and pentacyclic triterpanes from the Green River oil-shale of Eocene age (*285-287*) and an Early Precambrian shale (*285*). Tetraterpenes are also found in Green River oil-shale *288*).	Optically active triterpanes have been identified from a number of crude oils (*65, 81, 289*). Steranes have long been known in petroleum (*290*) and optically active varieties identified from some crude oils (*291*).
Porphyrins have been observed in a variety of rocks of many lithologies and all ages (*63, 293-298*).	Porphyrins in petroleum and asphalts were first attributed to biogenic sources in 1934. A large literature has built up over the years; only a few of the more recent references are cited here (*295, 298, 299, 300, 301*). Petroleum porphyrins are believed associated with protein fragments (*302*).

A wealth of laboratory data is now available on the low temperature pyrolysis of organic matter in recent sediments (*97–103*), ancient sedimentary rocks (*98, 100, 104–109*), on the kerogen from sedimentary rocks (*110–115*), and coals (*116*). These studies have demonstrated the importance of low temperature chemical reactions in the degradation of the organic matter in sediments and the subsequent generation of other compounds, some of which are not synthesized by living organisms. The role of sulfur in assisting the generation of some compounds has been noted (*117, 118*).

The laboratory studies on naturally occurring materials have been supplemented by maturation experiments using clean systems. The most relevant to the origin of the compounds in petroleum has been the work of Jurg and Eisma (*119–122*) on the conversion of fatty acids to normal paraffins. This experimental work supplements many theoretical and analytical studies (*123–130*) on the generation of *n*-paraffins from the abundant fatty acids present in living organisms (*131–134*) and found in recent sediments (*135–141*), ancient sedimentary rocks (*135, 142–146*), fossil algae (*147*), and petroleum (*146, 148*). However, this is a difficult chemical reaction, and probably the normal paraffins are related to the multi-ring naphthenes and multi-ring aromatics, as suggested by Spencer and Koons (*149*), with the fatty acids merely acting as solubilizers in the mobilization process.

Probably of equal importance to the observations of Erdman (*85*) cited above, and the pyrolysis experiments, is the large body of geological and geochemical evidence built up over many years' study of the world's sedimentary basins. The relation of crude oil and natural gas composition to position of the accumulations within the basins, especially with regard to depth and hence temperature, has been carefully documented. Only the more recent papers (*150–163*) need be cited as examples of the comprehensive studies undertaken. In the author's opinion, all this evidence indicates that maturation is by far the most important process governing the composition of crude oil (and natural gas) within the sedimentary environment. Indeed, it is only the presence of high energy barriers of up to 55 kcal/mole which protect many compounds (*8*) that allows the preservation of biological markers, thereby indicating the biological origin of petroleum.

Summarizing the maturation process we can observe that thermal influences commence with the initial compaction of the sediments, and in the early stages decarboxylation and hydrogenation are the main reactions. Continued burial increases thermal cracking, and true petroleum development commences—first with the heavy oils rich in ONS compounds. Deep burial causes increased splitting of C–C bonds, isotopic fractionation, and the development of many low molecular weight compounds, including natural gases. Paraffinic and aromatic crude oils characterize this phase. In the deepest strata only gas and condensates are found, and ultimately, possibly only graphitic materials. The upper temperature limits are probably 120°C for crude oils (25) and maybe 200°C for gas condensates. The actual depth at which the various reactions commence and cease will depend on the geothermal gradient (*156*). In general, minimum depths of 3000 feet of overburden are required before the generation of mobile petroleum takes place (25). Once mobile petroleum has been formed, it comes under the influence of the fluid pressure gradients already in existence in the sedimentary basin, and it is the migration and accumulation of petroleum (and natural gas) that we now consider.

Migration and Accumulation

Two important facts must be considered when we are concerned with the migration and accumulation of the fluid hydrocarbons (crude oil and natural gas) that have been generated from the immobile kerogenic material in the sedimentary rocks. First, the fluid hydrocarbons are an immiscible and minor part of the subsurface aqueous environment and must initially be mobilized so that they can move into this environment. Second, both the fluid hydrocarbons and the formation waters are subject to the same fluid pressure, thermal, and chemical gradients which exist in sedimentary basins. Porosities of the rocks within which the fluids move may range from over 50% in newly consolidated strata to less than 5% through compaction or reduction of porosity by mineral deposition from moving formation waters. Minerals commonly deposited include calcite, dolomite, silica, anhydrite, halite, and various clay minerals. The Western Canada sedimentary basin may be cited as an example of a large mature basin, and the calculated average basin porosity was 11.8% (*164*).

Sea water contains a wide variety of dissolved organic compounds, often at quite low concentrations (*165–171*), some of which are easily adsorbed onto suspended particulate matter (*172*). Undoubtedly some of this dissolved matter remains with the sea water as it is incorporated into the sedimentary environment. However, the relative amount of organic matter to inorganic matter (both water and mineral matter) in the sedimentary environment is considerably greater than in sea water, and this together with the reducing conditions present in the sediments enhances the presence of dissolved organic matter in the interstitial waters. Thus, the organic compounds found in formation waters include alkanes (*173–175*), naphthenic acids (*176, 177*), aromatic compounds

(*178–180*), amino acids (*181, 182*), and fatty acids (*135, 183*). Many of these compounds have surfactant properties and may play a significant role in the mobilization of petroleum.

The manner in which hydrocarbons are mobilized in the subsurface has been the subject of considerable controversy for many years, but there was little supporting experimental data. Baker (*184, 185*) carried out experiments on the enhanced solubility of hydrocarbons in water through the presence of selected surfactants. These led to his micellar solubility concept (*186–189*) in which he proposed that the hydrocarbons in the sedimentary rocks are mobilized into the aqueous phase in micellar solution. The hydrocarbons are then free to move with the water and are subject only to "unloading" by a variety of mechanisms. These include adsorption, temperature changes or salinity changes. Meinschein (*21*) has lucidly outlined the picture of the origin and accumulation of petroleum according to this concept. Additional experimental work supports these ideas (*173, 190, 191*). One particularly attractive feature of this concept is the postulation that it is the gathering of widely disseminated hydrocarbons that results in the phenomena of oil fields. This effectively eliminates the idea that the hydrocarbons in each oil field originated from a specific source rock. Also, the early development of the heavy, asphaltic oils of the "tar mats" which surround many sedimentary basins becomes clearer. These heavy oils are rich in ONS compounds, which are excellent surfactants, and would move most easily in the early stages of compaction when porosities and permeabilities were greater. Further, many of the tar mats are in the discharge regions of the basins, where the formation waters containing the hydrocarbons and surfactants come into contact with fresh (meteoric) waters, thereby initiating unloading.

Once the hydrocarbons have been "solubilized" in the formation water, they move with the water under the influence of elevation and pressure (fluid), thermal, electroosmotic and chemicoosmotic potentials. Of these, the fluid potential is the most important and the best known. The fluid potential is defined as the amount of work required to transport a unit mass of fluid from an arbitrary chosen datum (usually sea level) and state to the position and state of the point considered. The classic work of Hubbert (*192*) on the theory of groundwater motion was the first published account of the basinwide flow of fluids that considered the problem in exact mathematical terms as a steady-state phenomenon. His concept of formation fluid flow is shown in Figure 3A. However, incongruities in the relation between total hydraulic head and depth below surface in topographic low areas suggested that Hubbert's model was incomplete (*193*). Expanding on the work of Hubbert, Toth (*194, 195*) introduced a mathematical model in which exact flow patterns are obtained as solutions to formal boundary value problems. Specifically, his model (Figure 3B) was isotropic and homogeneous and, like Hubbert's, was based on the Laplace equation, but by superposition of a sinusoidal surface on the regional slope of the basin, Toth accounted for local topographic relief. Local changes in topography control an upper zone of a local fluid flow system, and this system is separated by an intermediate flow system from the main regional fluid flow regime which extends around the extremities of the model and is controlled by the regional topographic trend. Freeze and Witherspoon (*196–198*) pointed out a number of restrictions to the Toth model and developed both analytical and numerical methods of analysis (*199, 200*) for multilayer (nonhomogeneous), nonisotropic successions with a general configuration for the topographic surface and a sloping basement (Figure 3C).

The complex models generated by Freeze and Witherspoon are dimensionless, and thus the fluid flow nets apply to sedimentary basins of all sizes. The Toth-Freeze-Witherspoon model is meaningful when applied to an actual sedimentary basin, as exemplified by a recent study of the Western Canada sedimentary basin (*201, 202*). This is the only published study of a large sedimentary basin using the Toth-Freeze-Witherspoon model, and it shows that the main variables affecting the fluid potential distribution are topography and geology (lithology and permeability). The dominant fluid potential in any part of the basin corresponds closely to the fluid potential at the topographic surface in that part of the basin. Major recharge areas correspond to major upland areas, and major lowland regions are major discharge regions. Large rivers often exert significant drawdown effects (Figure 4). Relatively highly permeable beds, if sufficiently thick, can significantly affect the regional fluid potential distribution by drawing down the fluid potentials although the dominant features are topographically controlled. These observations apply to all sedimentary basins because of the dimensionless nature of the model. However, variations in flow pattern due to expulsion of formation water during compaction are not taken into account. Since compaction has not ceased in the Western Canada sedimentary basin, the effects of compaction are probably minimal—except possibly for very near surface sediments in relatively new basins. The model may also be used as an exploration tool in regions with the minimum of fluid pressure data (*203*).

An interesting implication of the Toth-Freeze-Witherspoon model is the deep penetration of surface water into the basin over geologic time. This has been used to derive a geochemical mass balance model for the mixing of surface water and diagenetically modified sea water in the Western Canada sedimentary basin using deuterium as a tracer (*204*).

The thermal potential is probably negligible as a force to move fluids in most sedimentary basins (*205*) but may be of significance in geothermal regions (*193*). Stallman (*206*) has assumed that movement of formation waters has an appreciable effect on the temperature distri-

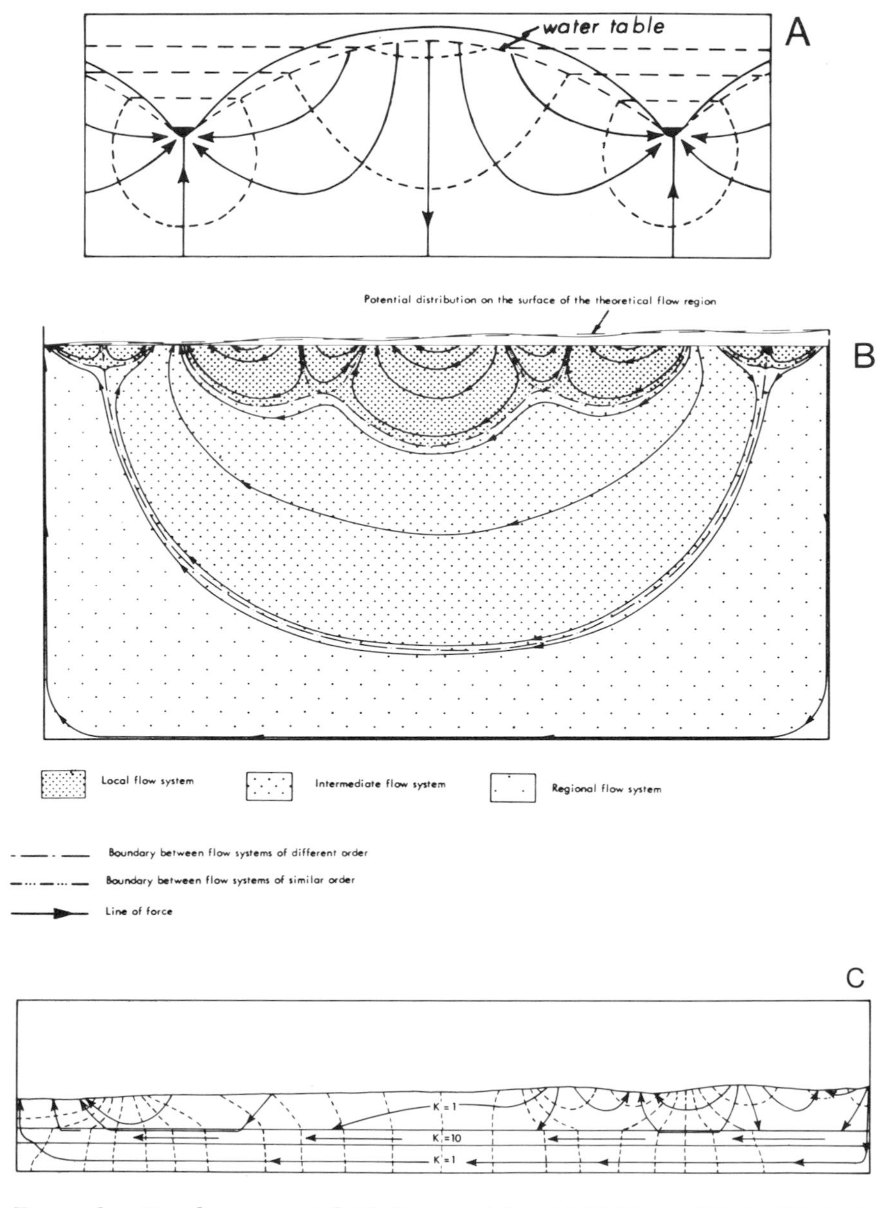

Figure 3. Development of fluid flow models: (A) Hubbert (192), *(B) Toth* (194, 195), *(C) Freeze and Witherspoon* (196–198)

bution in the subsurface. In the Western Canada sedimentary basin the separate pressure trends in the Upper Devonian Woodbend Group coral reefs described by Hitchon (*202*) are all associated with distinct temperature trends and correspond to marked differences in natural gas (*207–210*), crude oil (*153*), and formation water (*153*) composition. Such pressure, temperature, and fluid composition trends are related, in the case cited, to ease of access of the fluids in the reefs to the deep, hot, high potential fluids beneath the upland areas on the eastern flanks of the Rocky Mountains. Further, the degree to which the trends are sealed from cooler fluids above is important. The significance of the temperature gradient to the maturation process has been stressed by Philippi (*156*), and in part this accounts for the fluid compositional differences noted in the reefs. An ingenious example of the possible importance of thermal gradients in the subsurface is given in a recent paper by Mangelsdorf *et al.* (*211*) in which the authors calculate the effect of the migration of ions in formation waters in a thermal gradient. Such a mechanism does not explain the great enrichments in salinity found in many sedimentary basins but might be important in modifying salinity trends caused by other phenomena in regions of high thermal gradients. Watts (*212*) has suggested that thermal gradients could cause migration of petroleum through the effect of temperature on adsorption.

The electroosmotic and chemicoosmotic potentials have been evaluated as mechanisms causing fluid movement (*205*). The electroosmotic fluid movement is generated by electric potentials resulting from the existence of telluric currents and is assumed to have a negligible effect on fluid movement. The chemicoosmotic potential is important, however, in those sedimentary basins with both a wide range in composition of formation waters and shales which act as semipermeable membranes. A large literature has built up on the subject of the semipermeable membrane capacity of shales (*205, 213–228*), and it is now well documented that many fluid pressure anomalies, including closed fluid potential lows, in sedimentary basins owe their origin to chemicoosmotic forces (*202, 215, 219, 229*).

We may summarize the flow of fluids in sedimentary basins by noting that fluids in porous media move from regions of high energy to regions of low energy. The energy gradient to which the fluids react comprises an aggregate of potentials resulting from elevation and pressure, thermal, electroosmotic, and chemicoosmotic forces coupled in a system as yet incompletely described mathematically (*205*). The elevation and pressure potential can be completely described by the Toth-Freeze-Witherspoon model. It is only when the thermal, electroosmotic, and chemicoosmotic potentials influence the kinetic state of the fluid that their effect can be seen in terms of hydraulic head in the model, and then probably

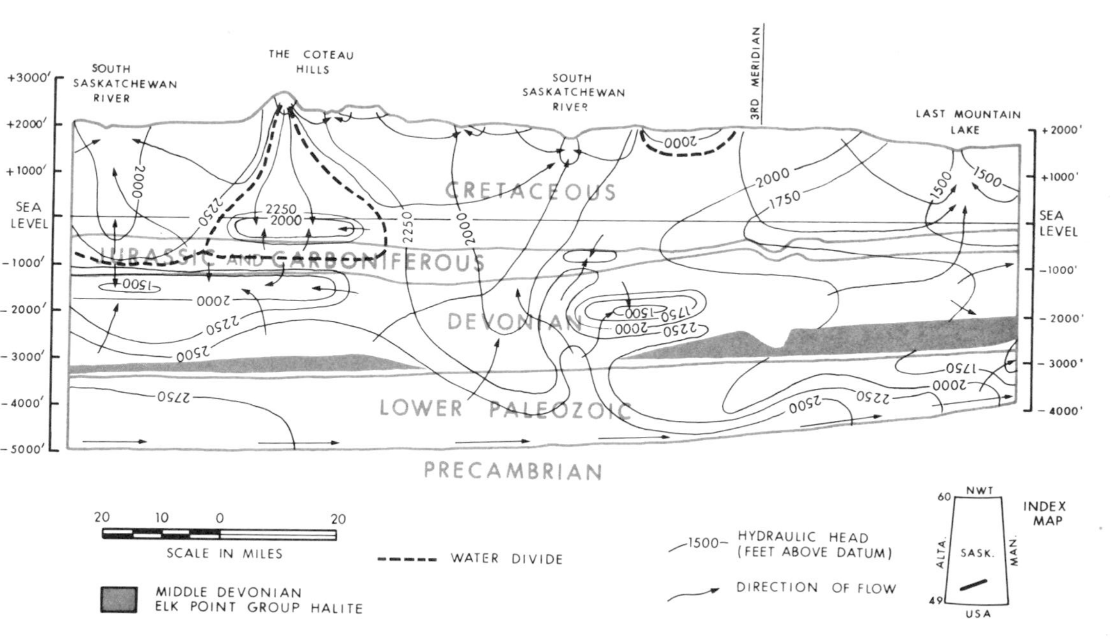

Figure 4. Hydraulic-head cross-section, southern Saskatchewan, Canada, showing the deep drawdown of the hydraulic head by the South Saskatchewan River. This has enabled fresh water to penetrate deep into the basin and has resulted in the solution of the Middle Devonian Elk Point group halite beneath the region of drawdown. The solution process has been assisted by the influx of fresh water in the region of the Cypress Hills (which lie southwest of the line of the cross-section) which has penetrated deep into the Devonian and is moving mainly into the line of the cross-section.

only a portion of that potential is observed. The thermal and electro-osmotic potentials probably contribute but a small fraction of the total energy of the system, although the former is the controlling factor in the maturation process. It is the chemico-osmotic potential that most strongly influences the state of movement of the fluid as described by the elevation and pressure potential in our model. The presence of an active chemico-osmotic potential manifests itself by apparently three-dimensionally closed hydraulic head lows, and by so-called anomalous pressures and salinities of formation waters.

The mobilized hydrocarbons, which move within the energy fields to which the formation water is subjected, may be unloaded by a variety of mechanisms, many of which are influenced by the water movement itself. For example, osmotic membrane effects result in salinity changes in the formation waters, which in turn can cause unloading of the hydrocarbons. The movement of formation water past bedded salt or anhydrite results in solution of the salt or anhydrite, increased salinity of the formation water and possible changes in the capacity of the formation water to accommodate hydrocarbons. Many similar situations can be visualized but because of the economic importance of studies of the unloading mechanism, no published reports are available.

Throughout the vicissitudes of mobilization, migration, and accumulation we have not discussed the trapping aspects of oilfields, which are emplaced in anticlinal, salt dome, stratigraphic, faulted or hydrodynamic situations, or combinations of these trapping classifications. In each of these situations which might appear conducive to oil accumulations the potential trap might be barren (water-filled) for a variety of causes, including the absence of potential source material; the presence of potential source material but lack of conditions of generation; the failure of migrating oil to reach the trap or escape of oil because of regional tilting or chemical dissipation; destruction of oil from weathering, bacteria, or extreme maturation (metamorphism); or, finally, because the trap was not formed until after the hydrocarbons had ceased migration. One further point that deserves mention is the relation of the trap to the trapping mechanism—for example, the movement of formation water through a sand pinch-out may cause salt-filtering and so induce disaccommodation of the migrating hydrocarbons. The relation of traps to trapping mechanisms deserves more careful thought than has been customary in the past.

Oil Fields and Ore Deposits

Oil fields and ore deposits in sedimentary rocks have several common features. Both are aggregates of widely dispersed matter accumulated at specific sites where physical and chemical changes in the aqueous carrier fluid caused unloading. These sites are often structurally controlled. Like oil fields, the fluid carrier for the ores is controlled within the sedimentary basin by the various potential fields described in the previous section. Neither the components of the ore body nor petroleum is usually very soluble. One major difference is the mobility of petroleum in its own right. However, both are subject to dissipation. The role of diagenetic processes in the formation of sedimentary mineral deposits has been reviewed by Amstutz and Bubenicek (*230*).

A good example of sedimentary ores which probably originated through interactions of formation waters are the lead-zinc-barite-fluorite deposits of Mississippi Valley type (*231–233*). The Pine Point lead-zinc deposits in the Northwest Territories, Canada have been studied with regard to their geochemistry and history (*234–238*). Most evidence points to an origin from formation waters. This viewpoint is supported by trace element studies of the present downdip formation waters (*239*) and examination of the broad picture of the flow and chemical composition of the formation waters (*153, 201, 202*). The detailed distribution of hydraulic-head in the Middle Devonian Keg River Formation of northern Alberta is shown in Figure 5. Fluid flow is generally updip, towards the northeast and the Pine Point lead-zinc deposit. Pine Point is situated at the outflow end of a porous and permeable limestone reef complex. The many oil pools in the Zama and Rainbow regions are emplaced in reef knolls (*see* diagrammatic cross-section in Figure 5) which lie upflow from Pine Point. The pressure depth chart in Figure 6 shows the relation of subsea elevation to pressure along this limestone reef complex (the line of open squares in Figures 5 and 6). The pressure gradient is greater than the hydrostatic gradient for fresh water, indicating updip flow—but is also close to the hydrostatic gradient for fresh water, indicating channel type flow and confirming the very porous and permeable nature of the limestone reef complex. The line of open circles in Figure 6 represents the pressure-depth relations along a N-S profile (see Figure 5). In the region where halite overlies the Keg River Formation, the formation pressures are about 400 psi greater than in the limestone reef channel, possibly because of reduction of permeability by salt deposition. To the north, along this line of section, pressures approach those in the channel. Thus the present hydraulic-head pattern suggests that both the Pine Point and Zama-Rainbow occurrences owe their origin to movement of formation waters updip along the porous and permeable limestone reef complex. This channel provided a suitable conduit for the fluids expelled from the dark shales present downdip. As they moved updip they deposited first oil and then the ores. Similar comprehensive attacks on the

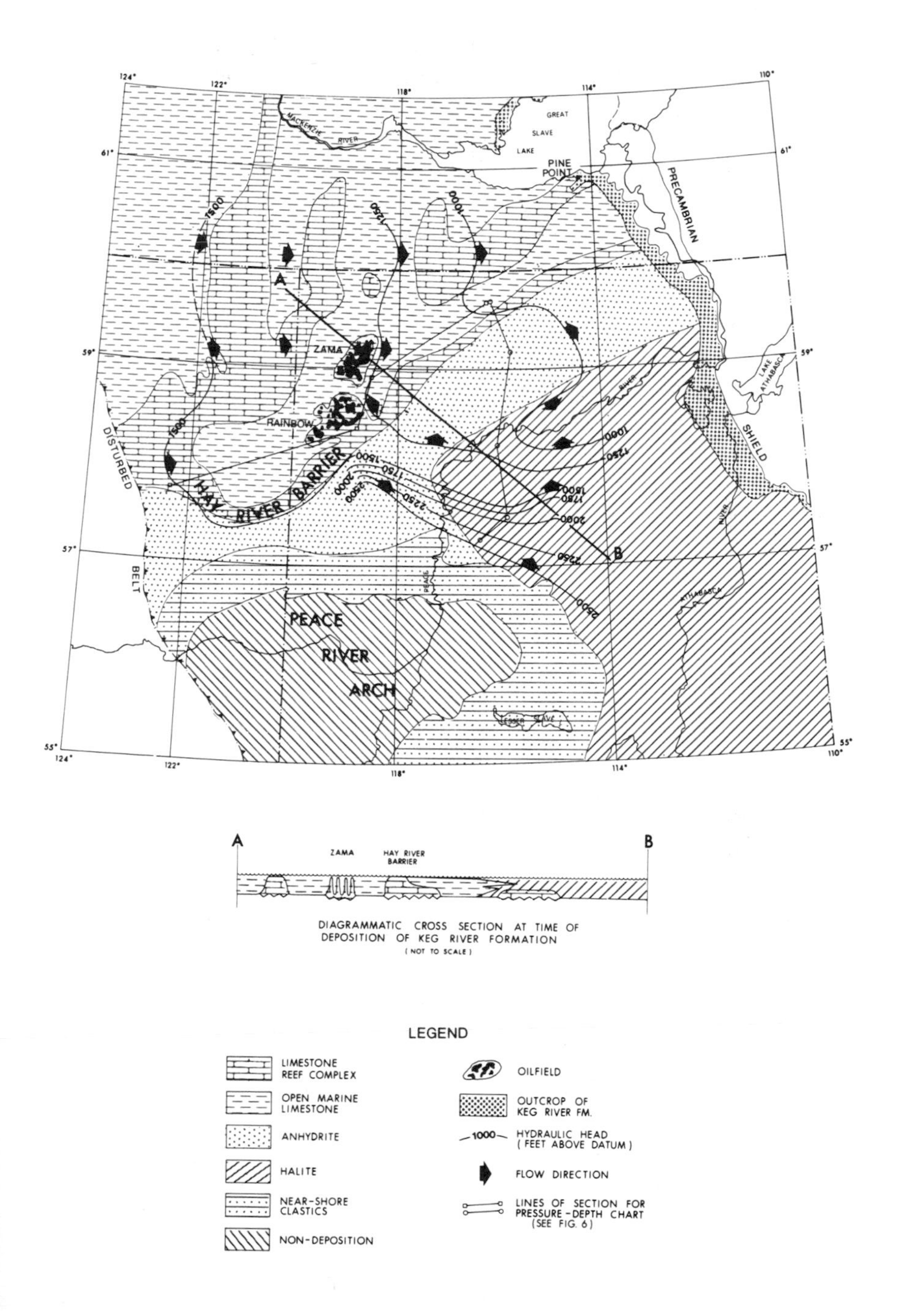

Figure 5. Hydraulic-head distribution in the Middle Devonian Keg River Formation of northern Alberta. Geology after McCamis and Griffith (303).

problems of the origins of other Mississippi Valley type sedimentary ore deposits would doubtless yield similar interesting results.

Saxby (*240*) has recently reviewed the entire field of metal-organic reactions in the geochemical cycle. The close relation of some oil fields and sedimentary ore deposits implies an intimate association of organic and inorganic components in formation waters. We have already discussed the effect of such broad relations as the association of salinity changes with hydrocarbon unloading. More subtle associations exist,

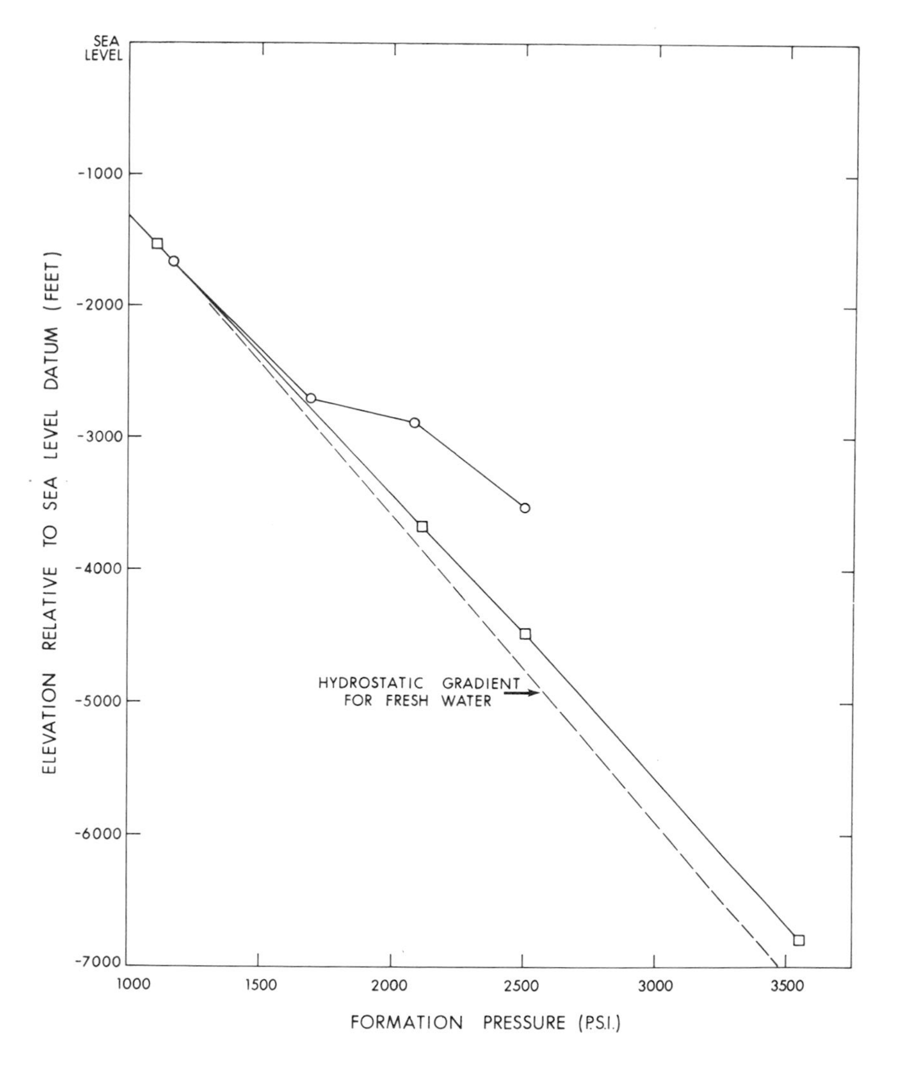

Figure 6. Pressure-depth chart for the Middle Devonian Keg River Formation of northern Alberta. See *Figure 5 for lines of section.*

however. One of the best documented is the demetallation of chlorins in recent sediments and the subsequent introduction of nickel, iron, or vanadyl to produce the metal-complexed porphyrins of ancient sedimentary rocks and petroleum. The asphaltenes also complex heavy metals, particularly the vanadyl complex (*241*), and the distribution of metals in the sedimentary cycle may be influenced by components like amino acids which act as metal binders (*242*). Chelation (metal-organic complex formation) in the geological environment has been studied by a number of workers (*243–246*) and has been invoked as a means of transporting complex compounds (*247, 248*). Solomon (*249*) has suggested that clay minerals may act as electron donors or acceptors in organic reactions. In summary, the mode of transportation during genesis of sedimentary ores is a fascinating subject for research, and its study should provide valuable information on organic-inorganic reactions in the geological environment.

Discussion

It is difficult not to be categorical when writing a review because of the desire to present a comprehensive, coherent analysis of the subject being considered. There are many aspects of the origin of oil that we do not understand, but I believe that a comparison of this review and its companion review (*8*) with a review by the same authors some five years ago (*9*) will indicate the advances made in this short time. They have been concerned with the chemistry of the components in crude oils, particularly with those compounds now recognized as biological markers. Of equal importance has been the advancement of our understanding of fluid flow to the extent where we can now visualize some of the extreme complexity of subsurface flow patterns. From a practical point of view it is more likely that continuing studies of fluid flow and formation waters will be of value in the exploration for petroleum although advances in the chemical typing of oil and sedimentary rocks may play their part in exploration. However, we still do not know how oil is mobilized. Perhaps it is not important to understand this intermediate step between source and reservoir, but it seems to this reviewer that a better understanding of the process of mobilization would assist the application of research projects in both organic geochemistry and fluid flow to the exploration for petroleum.

Summary

The geological constraints on the origin of oil are broad and minimal—namely, generation of a mobile phase from organic matter in the sedimentary environment through low temperature reactions and emplacement within that sedimentary environment (or subjacent igneous or metamorphic rocks) in situations where the geochemical characteristics of the aqueous carrier fluid cause unloading of the mobile organic phase to form oil fields. The vicissitudes of structural and stratigraphic variations within the sedimentary environment, and the changes of permeability and porosity with time, are all factors complexing an otherwise relatively simple situation—they are not constraints, as the number of dry structural and stratigraphic "traps" testifies. Our understanding of the origin of oil will be enhanced only through comprehensive research programs which are carried out on the premise that we are concerned both with a previously mobile organic phase, generated through low temperature reactions, and with an aqueous carrier fluid whose chemical composition and flow characteristics are related to the chemistry and physics of the sedimentary environment within which both are found.

Literature Cited

(1) Russell, W. L., "Principles of Petroleum Geology," McGraw-Hill, New York, 1951.
(2) Hobson, G. D., "Some Fundamentals of Petroleum Geology," Oxford University Press, London, 1954.
(3) Landes, K. K., "Petroleum Geology," Wiley, New York, 1959.
(4) Moody, G. B., Ed., "Petroleum Exploration Handbook," McGraw-Hill, New York, 1961.
(5) Levorsen, A. I., "Geology of Petroleum," Freeman, San Francisco, 1967.
(6) Louis, M., "Cours de Géochimie du Pétrole," Technip, Paris, 1967.
(7) Dott, R. H., Reynolds, M. J., "Sourcebook for Petroleum Geology," American Association of Petroleum Geologists, Memoir 5, 1969.
(8) Hodgson, G. W., "Origin of Oil: Chemical Contraints," ADVAN. CHEM. SER. (1971) **103**, 1.
(9) Hodgson, G. W., Hitchon, B., "Research Trends in Petroleum Genesis," *Proc. Commonwealth Mining Metallurgical Congr., 8th, Australia and New Zealand* (1965) **5**, 9–19.
(10) Breger, I. A., Ed., "Organic Geochemistry," Pergamon, Oxford, 1963.
(11) Colombo, U., Hobson, G. D., Eds., "Advances in Organic Geochemistry," Macmillan, New York, 1964.
(12) Hobson, G. D., Louis, M. C., Eds., "Advances in Organic Geochemistry," Pergamon, Oxford, 1966.
(13) Nagy, B., Colombo, U., Eds., "Fundamental Aspects of Petroleum Geochemistry," Elsevier, Amsterdam, 1967.
(14) Eglinton, G., Murphy, M. T. J., Eds., "Organic Geochemistry," Springer-Verlag, Berlin, 1969.
(15) Schenck, P. A., Havenaar, I., Eds., "Advances in Organic Geochemistry," Pergamon, Oxford, 1969.
(16) Hobson, G. D., Speers, G. C., Eds., "Advances in Organic Geochemistry," Pergamon, Oxford, 1970.
(17) Andreev, P. F., Bogomolov, A. I., Dobryanskii, A. F., Kartsev, A. A., "Transformation of Petroleum in Nature," Pergamon, Oxford, 1968.
(18) Manskaya, S. M., Drozdova, T. V., "Geochemistry of Organic Substances," Pergamon, Oxford, 1968.
(19) Davis, J. B., "Petroleum Microbiology," Elsevier, Amsterdam, 1967.

(20) Stevens, N. P., "Origin of Petroleum—A Review," *Am. Assoc. Petrol. Geol. Bull.* (1956) **40** (1), 51–61.
(21) Meinschein, W. G., "Origin of Petroleum," *Am. Assoc. Petrol. Geol. Bull.* (1959) **43** (5), 925–943.
(22) Meinschein, W. G., "Origins of Natural Gas and Petroleum," *Nature* (1968) **220** (5173), 1185–1189.
(23) Krejci-Graf, K., "Origin of Oil," *Geophys. Prospecting* (1963) **11** (3), 244–275.
(24) Erdman, J. G., "Geochemistry of the High Molecular Weight Non-hydrocarbon Fraction of Petroleum," in "Advances in Organic Geochemistry," pp. 215–237, Macmillan, New York, 1964.
(25) Hedberg, H. D., "Geologic Aspects of Origin of Petroleum," *Am. Assoc. Petrol. Geol. Bull.* (1964) **48** (11), 1755–1803.
(26) Hedberg, H. D., "Geologic Controls on Petroleum Genesis," *Proc. World Petrol. Congr., 7th* (1967) **2** (1), 3–11.
(27) Hobson, G. D., "The Organic Geochemistry of Petroleum," *Earth-Sci. Rev.* (1966) **2,** 257–276.
(28) Smith, H. M., "Crude Oil: Qualitative and Quantitative Aspects," *U.S. Bur. Mines Inf. Circ.* (1966) **8286.**
(29) Bestougeff, M. A., "Petroleum Hydrocarbons," in "Fundamental Aspects of Petroleum Geochemistry," pp. 77–103, Elsevier, Amsterdam, 1967.
(30) Colombo, U., "Origin and Evolution of Petroleum," in "Fundamental Aspects of Petroleum Geochemistry," pp. 331–369, Elsevier, Amsterdam, 1967.
(31) Costantinides, G., Arich, G., "Non-hydrocarbon Compounds in Petroleum," in "Fundamental Aspects of Petroleum Geochemistry," pp. 109–175, Elsevier, Amsterdam, 1967.
(32) Degens, E. T., "Diagenesis of Organic Matter," in "Diagenesis in Sediments," pp. 343–390, Elsevier, Amsterdam, 1967.
(33) Ellison, S. P., "Petroleum (origin)," Encyclopedia of Chemical Technology, Vol. 14, pp. 838–845, Interscience, New York, 1967.
(34) McCarthy, E. D., "Treatise on Organic Geochemistry," *Univ. Calif. (Berkeley) Rept.* (1967) No. **UCRL-17758.**
(35) McCarthy, E. D., Calvin, M., "Organic Geochemical Studies. I. Molecular Criteria for Hydrocarbon Genesis," *Nature* (1967) **216** (5116), 642–647.
(36) Vassoyevich, N. B., Amosov, G. A., "Geological and Geochemical Evidence of Origin of Crude Oil from Living Matter," *Proc. U.S.S.R. All-Union Oil Gas Genesis Symp., Moscow* (1967), 5–22.
(37) Staplin, F. L., "Sedimentary Organic Matter, Organic Metamorphism, and Oil and Gas Occurrence," *Bull. Can. Petrol. Geol.* (1969) **17** (1), 47–66.
(38) Hills, I. R., Smith, G. W., Whitehead, E. V., "Hydrocarbons from Fossil Fuels and Their Relationship with Living Organisms," *J. Inst. Petrol.* (1970) **56** (549), 127–137.
(39) Vassoyevich, N. B., Korchagina, Y. I., Lopatin, N. V., Chernyshev, V. V., "Principal Phase of Oil Formation," *Internt. Geol. Rev.* (1970) **12** (11), 1276–1296.
(40) Cutbill, J. L., Funnell, B. M., "Numerical Analysis of The Fossil Record," in "The Fossil Record," pp. 791–820, Geological Society of London, 1967.
(41) Knebel, G. M., Rodriguez-Eraso, G., "Habitat of Some Oil," *Am. Assoc. Petrol. Geol. Bull.* (1956) **40** (4), 547–561.
(42) Weeks, L. G., "Origin, Migration, and Occurrence of Petroleum," in "Petroleum Exploration Handbook," pp. 5-1 to 5-50, McGraw-Hill, New York, 1961.

(43) Odum, H. T., Hoskin, C. M., *Univ. Texas Inst. Marine Sci.* (1958) **5,** 16–46.
(44) Cox, B. B., "Transformation of Organic Material into Petroleum under Geological Conditions, The Geological Fence," *Am. Assoc. Petrol. Geol. Bull.* (1946) **30** (5), 645–659.
(45) Weeks, L. G., "Habitat of Oil and Some Factors that Control It," in "Habitat of Oil," pp. 1–61, American Association of Petroleum Geologists, Tulsa, 1958.
(46) Eglinton, G., "Hydrocarbons and Fatty Acids in Living Organisms and Recent and Ancient Sediments," in "Advances in Organic Geochemistry," pp. 1–24, Pergamon, Oxford, 1969.
(47) Belsky, T., "Chemical Evolution and Organic Geochemistry," Ph.D. Thesis, University of California, Berkeley, 1966.
(48) Degens, E. T., Schmidt, H., "Paleobiochemistry, A New Active Field in the Study of Evolution," *Palaeontol. Z.* (1966) **40** (3–4), 218–229.
(49) Eglinton, G., "Recent Advances in Organic Geochemistry," *Geol. Rundsch.* (1966) **55** (3), 551–567.
(50) Eglinton, G., Calvin, M., "Chemical Fossils," *Sci. Am.* (1967) **216** (1), 32–43.
(51) Vogel, H. J., Vogel, R. H., "Some Chemical Glimpses of Evolution," *Chem. Eng. News* (1967) **45** (52), 90–97.
(52) Calvin, M., "Molecular Paleontology," *Perspectives Biol. Med.* (1969) **13** (1), 45–62.
(53) Welte, D. H., "Determination of $^{13}C/^{12}C$ Isotope Ratios of Individual Higher *n*-Paraffins from Different Petroleums," in "Advances in Organic Geochemistry," pp. 269–277, Pergamon, Oxford, 1969.
(54) Silverman, S. R., "Investigations of Petroleum Origin and Evolution Mechanisms by Carbon Isotope Studies," in "Isotopic and Cosmic Chemistry," pp. 92–102, North-Holland Publishing Co., Amsterdam, 1964.
(55) Silverman, S. R., "Carbon Isotopic Evidence for the Role of Lipids in Petroleum Formation," *J. Am. Oil Chemists Soc.* (1966) **44,** 691–695.
(56) Craig, H., "The Geochemistry of the Stable Carbon Isotopes," *Geochim. Cosmochim. Acta* (1953) **3** (2/3), 53–92.
(57) Silverman, S. R., Epstein, S., "Carbon Isotopic Compositions of Petroleums and Other Sedimentary Organic Materials," *Am. Assoc. Petrol. Geol. Bull.* (1958) **42** (5), 998–1012.
(58) Tarlo, L. B. H., "Biochemical Evolution and the Fossil Record," in "The Fossil Record," pp. 119–132, Geological Society of London, 1967.
(59) Baldwin, E., "An Introduction to Comparative Biochemistry," Cambridge University Press, Cambridge, 1949.
(60) Florkin, M., "Biochemical Evolution," Academic, New York, 1949.
(61) Florkin, M., Mason, H. S., "Comparative Biochemistry—A Comprehensive Treatise," Vols. I–VII, Academic, New York, 1963.
(62) Dunning, H. N., "Geochemistry of Organic Pigments," in "Organic Geochemistry," pp. 367–430, Pergamon, Oxford, 1963.
(63) Hodgson, G. W., Hitchon, B., Taguchi, K., Baker, B. L., Peake, E., "Geochemistry of Porphyrins, Chlorins and Polycyclic Aromatics in Soils, Sediments and Sedimentary Rocks," *Geochim. Cosmochim. Acta* (1968) **32** (7), 737–772.
(64) Hodgson, G. W., Baker, B. L., Peake, E., "Geochemistry of Porphyrins," in "Fundamental Aspects of Petroleum Geochemistry," pp. 177–259, Elsevier, Amsterdam, 1967.
(65) Hills, I. R., Whitehead, E. V., "Pentacyclic Triterpanes from Petroleum and Their Significance," in "Advances in Organic Geochemistry," pp. 89–110, Pergamon, Oxford, 1970.

(66) Degens, E. T., Spencer, D. W., Parker, R. H., "Paleobiogeochemistry of Molluscan Shell Proteins," *Comp. Biochem. Physiol.* (1967) **20,** 553–579.
(67) Ghiselin, M. T., Degens, E. T., Spencer, D. W., Parker, R. H., "A Phylogenetic Survey of Molluscan Shell Matrix Proteins," *Breviora* (1967) **262,** 1–35.
(68) Hotta, N., "Amino Acids in Fossil Shells of *Glycymeris* from Japan," *J. Geol. Soc. Japan* (1965) **71** (842), 554–566.
(69) Hotta, N., "Residual Amino Acids in the Shells of *Turritella* from Japan," *J. Geol. Soc. Japan* (1967) **73** (7), 315–324.
(70) Degens, E. T., Love, S., "Comparative Studies of Amino Acids in Shell Structures of *Gyraulus trochiformis,* Stahl, from the Tertiary of Steinheim, Germany," *Nature* (1965) **205** (4974), 876–878.
(71) Swain, F. M., Kraemer, S. A., "Amino Acid Components of Some Paleozoic Plant Fossils and Rock Samples," *J. Paleontol.* (1969) **43** (2), 546–550.
(72) Swain, F. M., Pakalus, G. V., Bratt, J. G., "Possible Taxonomic Interpretation of Some Palaeozoic and Precambrian Carbohydrate Residues," in "Advances in Organic Geochemistry," pp. 469–491, Pergamon, Oxford, 1970.
(73) Oró, J., Laseter, J. L., Weber, D., "Alkanes in Fungal Spores," *Science* (1966) **154** (3747), 399–400.
(74) Clark, R. C., Blumer, M., "Distribution of *n*-Paraffins in Marine Organisms and Sediment," *Limnol. Oceanog.* (1967) **12** (1), 79–87.
(75) Han, J., Calvin, M., "Hydrocarbon Distribution of Algae and Bacteria and Microbiological Activity in Sediments," *Proc. Natl. Acad. Sci.* (1969) **64** (2), 436–443.
(76) Holton, R. W., Blecker, H. H., Stevens, T. S., "Fatty Acids in Blue-Green Algae: Possible Relation to Phylogenetic Position," *Science* (1968) **160** (3827), 545–547.
(77) Hedberg, H. D., "Significance of High-Wax Oils with Respect to Genesis of Petroleum," *Am. Assoc. Petrol. Geol. Bull.* (1968) **52** (5), 736–750.
(78) Biederman, E. W., "Significance of High-Wax Oils with Respect to Genesis of Petroleum: Commentary," *Am. Assoc. Petrol. Geol. Bull.* (1969) **53** (7), 1500–1502.
(79) Reed, K. J., "Environment of Deposition of Source Beds of High-Wax Oil," *Am. Assoc. Petrol. Geol. Bull.* (1969) **53** (7), 1502–1506.
(80) Brooks, J. D., Gould, K., Smith, J. W., "Isoprenoid Hydrocarbons in Coal and Petroleum," *Nature* (1969) **222** (5190), 257–259.
(81) Hills, I. R., Whitehead, E. V., "Triterpanes in Optically Active Petroleum Distillates," *Nature* (1966) **209** (5027), 977–979.
(82) Robinson, R., "Duplex Origin of Petroleum," *Nature* (1963) **199** (4889), 113–114.
(83) Rudakov, G., "Recent Developments in the Theory of the Non-Biogenic Origin of Petroleum," *Chem. Geol.* (1967) **2,** 179–185.
(84) Rudakow (sic) G. W., "The Formation of Petroleum in Depth—Some New Aspects of this Theory," *Erdöl Kohle* (1970) **23** (7), 404–410.
(85) Erdman, J. G., "Petroleum—Its Origin in the Earth," in "Fluids in Subsurface Environments," *Am. Assoc. Petrol. Geol. Mem.* (1964) **4,** 20–52.
(86) Erdman, J. G., "Geochemical Origins of the Low Molecular Weight Hydrocarbon Constituents of Petroleum and Natural Gases," *Proc. World Petrol. Congr., 7th* (1967) **2,** 13–24.
(87) Veber, V. V., Turkel'taub, N. M., "Gaseous Hydrocarbons in Recent Sediments," *Geol. Neft. Gaz.* (1958) **2** (8), 39–44.
(88) Veber, V. V., Turkel'taub, N. M., "Formation of Gaseous Hydrocarbons Depending on Facies Depositions," *Geol. Neft. Gaz.* (1965) **8,** 41–48.

(89) Dunton, M. L., Hunt, J. M., "Distribution of Low Molecular-Weight Hydrocarbons in Recent and Ancient Sediments," *Am. Assoc. Petrol. Geol. Bull.* (1962) **46** (12), 2246–2248.

(90) Thompson, R. R., Creath, W. B., "Low Molecular Weight Hydrocarbons in Recent and Fossil Shells," *Geochim. Cosmochim. Acta* (1966) **30** (11), 1137–1152.

(91) Rogers, M. A., Koons, C. B., "Generation of Light Hydrocarbons and Establishment of Normal Paraffin Preferences in Crude Oils," ADVAN. CHEM. SER. (1971) **103,** 67.

(92) Louis, M., "Essais sur l'Evolution de Petrole a Faible Temperature en Presence de Mineraux," in "Advances in Organic Geochemistry," pp. 261–278, Pergamon, Oxford, 1966.

(93) Henderson, W., Eglinton, G., Simmonds, P., Lovelock, J. E., "Thermal Alteration as a Contributory Process to the Genesis of Petroleum," *Nature* (1968) **219,** 1012–1016.

(94) Galwey, A. K., "Reactions of Alcohols Adsorbed on Montmorillonite and the Role of Minerals in Petroleum Genesis," *Chem. Commun.* (1969), 577–578.

(95) Galwey, A. K., "Heterogeneous Reactions in Petroleum Genesis and Maturation," *Nature* (1969) **223** (5212), 1257–1260.

(96) Califet, Y., Oudin, J. L., "Influence of Temperature, Pressure, and a Clay Mineral on the Evolution of the Chemical Structure of an Aromatic Fraction of a Crude Oil," in "Advances in Organic Geochemistry," p. 153, Pergamon, Oxford, 1970.

(97) Mulik, J. D., Erdman, J. G., "Genesis of Hydrocarbons of Low Molecular Weight in Organic-Rich Aquatic Systems," *Science* (1963) **141** (3583), 806–807.

(98) Hoering, T. C., Abelson, P. H., "Hydrocarbons From the Low-Temperature Heating of Kerogen," *Carnegie Inst. Washington Year Book 63, Papers Geophys. Lab.* (1964) **1440,** 256–258.

(99) Califet, Y., Louis, M., "Contribution to the Knowledge Concerning the Stability of Amino Acids Contained in Sedimentary Rocks," *Compt. Rend.* (1965) **261** (18), 3645–3646.

(100) Sellers, G. A., "Hydrothermal Experiments on the Thermal Stability of Amino Substances in Sediments," Ph.D. Thesis (1966), California Institute of Technology.

(101) Weide, B. Van der, "Evolution des *n*-Paraffines par Traitement Thermique de Sédiments Marins Recents," *Bull. Centre Rech. PAU-SNPA* (1967) **1** (1), 161–164.

(102) Mitterer, R. M., Hoering, T. C., "Production of Hydrocarbons from the Organic Matter in a Recent Sediment," *Carnegie Inst. Washington Year Book 66, Papers Geophys. Lab.* (1968) **1499,** 510–514.

(103) Hoering, T. C., "Reactions of the Organic Matter in a Recent Marine Sediment," *Carnegie Inst. Washington Year Book 67, Papers Geophys. Lab.* (1969) **1520,** 199–201.

(104) Welte, D. H., "Hydrocarbon Genesis in Sedimentary Rocks—Investigation of the Thermal Decomposition of Kerogen with Particular Consideration of the *n*-Paraffin Genesis," *Geol. Rundsch.* (1966) **55** (1), 131–144.

(105) Louis, M. C., "Influence of Temperature and Depth on Formation of Hydrocarbons, Particularly in Kerogen Shales," *Proc. World Petrol. Congr., 7th* (1967) **2** (1), 47–60.

(106) Gaertner, H. R. von, Kroepelin, H., "A Study of the Diagenetic Evolution of the Organic Substances in Bituminous Shales as Illustrated by the Posidonia Shale," *Beih. Geol. Jahrbuch* (1968) **58,** 1–12.

(107) Kroepelin, H., "Chemical Study of the Northwest Germany Posidonia Shales and Their Organic Substances," *Beih. Geol. Jahrbuch* (1968) **58,** 499–563.
(108) Douglas, A. G., Eglinton, G., Henderson, W., "Thermal Alteration of the Organic Matter in Sediments," in "Advances in Organic Geochemistry," pp. 369–388, Pergamon, Oxford, 1970.
(109) Giraud, A., "Application of Pyrolysis and Gas Chromatography to Geochemical Characterization of Kerogen in Sedimentary Rock," *Am. Assoc. Petrol. Geol. Bull.* (1970) **54** (3), 439–455.
(110) Robinson, W. E., Lawlor, D. L., "Constitution of Hydrocarbon-Like Materials Derived from Kerogen Oxidation Products," *Fuel* (1961) **40** (5), 375–388.
(111) Hoering, T. C., Abelson, P. H., "Hydrocarbons from Kerogen," *Carnegie Inst. Washington Year Book 62, Papers Geophys. Lab.* (1963) **1412,** 229–234.
(112) Robinson, W. E., Lawlor, D. L., Cummins, J. J., Fester, J. I., "Oxidation of Colorado Oil Shale," *U.S. Bur. Mines, Rept. Invest.* (1963) **6166.**
(113) McIver, R. D., "Composition of Kerogen-Clue to Its Role in the Origin of Petroleum," *Proc. World Petrol. Congr., 7th* (1967) **2** (1), 25–36.
(114) Robinson, W. E., Dinneen, G. U., "Constitutional Aspects of Oil-Shale Kerogen," *Proc. World Petrol. Congr., 7th* (1967) **2,** Paper PD-14.
(115) Bordenave, M., Combaz, A., Giraud, A., "Influence de l'Origine des Matières Organiques et de leur Degré d'Evolution sur les Produits de Pyrolyse du Kérogène," in "Advances in Organic Geochemistry," pp. 389–405, Pergamon, Oxford, 1970.
(116) Brooks, J. D., Smith, J. W., "The Diagenesis of Plant Lipids During the Formation of Coal, Petroleum and Natural Gas—11. Coalification and the Formation of Oil and Gas in the Gippsland Basin," *Geochim. Cosmochim. Acta* (1969) **33** (10), 1183–1194.
(117) Douglas, A. G., Mair, B. J., "Sulfur: Role in Genesis of Petroleum," *Science* (1965) **147** (3657), 499–501.
(118) Guseva, A. N., Faingersh, L. A., Chakhmakhohev, V. A., "Changes in Composition of Oils Due to Sulfurization," *Izv. Akad. Nauk SSSR, Ser. Geol.* (1968) **6,** 67–73.
(119) Jurg, J. W., Eisma, E., "Petroleum Hydrocarbons: Generation from Fatty Acid," *Science* (1964) **144** (3625), 1451-1452.
(120) Jurg, J. W., Eisma, E., "The Mechanism of the Generation of Petroleum Hydrocarbons from a Fatty Acid," in "Advances in Organic Geochemistry," pp. 367–368, Pergamon, Oxford, 1970.
(121) Eisma, E., Jurg, J. W., "Fundamental Aspects of the Diagenesis of Organic Matter and the Formation of Hydrocarbons," *Proc. World Petrol. Congr., 7th* (1967) **2,** 61–72.
(122) Eisma, E., Jurg, J. W., "Fundamental Aspects of the Generation of Petroleum," in "Organic Geochemistry," pp. 676–698, Springer-Verlag, Berlin, 1969.
(123) Cooper, J. E., Bray, E. E., "A Postulated Role of Fatty Acids in Petroleum Formation," *Geochim. Cosmochim. Acta* (1963) **27** (11), 1113–1127.
(124) Mair, B. J., "Terpenoids, Fatty Acids and Alcohols as Source Materials for Petroleum Hydrocarbons," *Geochim. Cosmochim. Acta* (1964) **28** (8), 1303–1321.
(125) Kvenvolden, K. A., "Normal Fatty Acids in Sediments," *J. Am. Oil Chemists Soc.* (1967) **44,** 628–636.
(126) Kvenvolden, K. A., "Geochemical Exploration," U. S. Patent **3,480,396** (1969).

(127) Kvenvolden, K. A., "Evidence for Transformations of Normal Fatty Acids in Sediments," in "Advances in Organic Geochemistry," pp. 335–366, Pergamon, Oxford, 1970.
(128) Kvenvolden, K. A., Weiser, D., "A Mathematical Model of a Geochemical Process: Normal Paraffin Formation from Normal Fatty Acid," *Geochim. Cosmochim. Acta* (1967) **31** (8), 1281–1309.
(129) Maclean, I., Eglinton, G., Douraghi-Zadeh, K., Ackman, R. G., Hooper, S. N., "Correlation of Stereoisomerism in Present Day and Geologically Ancient Isoprenoid Fatty Acids," *Nature* (1968) **218** (5146), 1019–1024.
(130) Parker, P. L., "Fatty Acids and Alcohols," in "Organic Geochemistry," pp. 357–373, Springer-Verlag, Berlin, 1969.
(131) Parker, P. L., Leo, R. F., "Fatty Acids in Blue-Green Algal Mat Communities," *Science* (1965) **148** (3668), 373–374.
(132) Collier, A., "Fatty Acids in Certain Plankton Organisms," in "Estuaries," *Am. Assoc. Advan. Sci., Publ.* (1967) **83,** 353–360.
(133) Maurer, L. G., Parker, P. L., "Fatty Acids in Sea Grasses and Marsh Plants," *Contrib. Marine Sci.* (1967) **12,** 113–119.
(134) Parker, P. L., Van Baalen, C., Maurer, L., "Fatty Acids in Eleven Species of Blue-Green Algae: Geochemical Significance," *Science* (1967) **155** (3763), 707–708.
(135) Cooper, J. E., "Fatty Acids in Recent and Ancient Sediments and Petroleum Reservoir Waters," *Nature* (1962) **193** (4817), 744–746.
(136) Abelson, P. H., Hoering, T. C., Parker, P. L., "Fatty Acids in Sedimentary Rocks," in "Advances in Organic Geochemistry," pp. 169–174, Pergamon, Oxford, 1964.
(137) Leo, R. F., Parker, P. L., "Branched-Chain Fatty Acids in Sediments," *Science* (1966) **152** (3722), 649–650.
(138) Parker, P. L., "Fatty Acids in Recent Sediment," *Contrib. Marine Sci.* (1967) **12,** 132–142.
(139) Peterson, D. H., "Fatty Acid Composition of Certain Shallow-Water Marine Sediments," Ph.D. Thesis (1967), Washington University, Seattle.
(140) Cooper, W. J., Blumer, M., "Linear, Iso and Anteiso Fatty Acids in Recent Sediments of the North Atlantic," *Deep-sea Res.* (1968) **15** (5), 535–540.
(141) Hoering, T. C., "Branched-Chain Fatty Acids in Recent Sediments," *Carnegie Inst. Washington Year Book 67, Papers Geophys. Lab.* (1969) **1520,** 201–202.
(142) Kvenvolden, K. A., "Molecular Distributions of Normal Fatty Acids and Paraffins in Some Lower Cretaceous Sediments," *Nature* (1966) **209** (5023), 573–577.
(143) Burlingame, A. L., Simoneit, B. R., "Isoprenoid Fatty Acids Isolated From the Kerogen Matrix of the Green River Formation (Ecocene)," *Science* (1968) **160** (3827), 531–533.
(144) Welte, D. H., Ebhardt, G., "Distribution of Long Chain *n*-Paraffins and *n*-Fatty Acids in Sediments From the Persian Gulf," *Geochim. Cosmochim. Acta* (1968) **32** (4), 465–466.
(145) Han, J., Calvin, M., "Occurrence of Fatty Acids and Aliphatic Hydrocarbons in a 3.4 Billion-Year-Old Sediment," *Nature* (1969) **224** (5219), 576–577.
(146) Van Hoeven, W., Maxwell, J. R., Calvin, M., "Fatty Acids and Hydrocarbons As Evidence of Life Processes in Ancient Sediments and Crude Oils," *Geochim. Cosmochim Acta* (1969) **33** (7), 877–881.
(147) Das, S. K., Smith, E. D., "Fatty Acids in Fossil Algae of Different Geologic Ages," *N. Y. Acad. Sci. Annals* (1968) **147** (10), 413–418.

(148) Glogoszowksi, J. J., Sliwiok, J., "Fatty Acids in Petroleum and in Dispersed Organic Substance," *Z. Angew. Geol.* (1969) **15** (10), 523–527.
(149) Spencer, D. W., Koons, C. B., "Studies on Origin of Crude Oil: Statistical Analyses of Crude Oil Data," *Abst., Am. Assoc. Petrol. Geol. Bull.* (1970) **54** (5), 871.
(150) Andreev, P. F., "On Geochemical Transformations of Oil in the Lithosphere," *Geokhimiya* (1962) **10,** 880–889.
(151) McIver, R. D., "The Crude Oils of Wyoming—Product of Depositional Environment and Alteration," *Ann. Field Conf. Wyoming Geol. Assoc., 17th, Guidebook* (1962), 248–251.
(152) Dobryansky, A. F., "La Transformation du Pétrole Brut dans la Nature," *Rev. Inst. Franc. Pétrole* (1963) **18** (1), 41–49.
(153) Hitchon, B., "Formation Fluids," in "Geological History of Western Canada," Chap. 15, pp. 201–217, Alberta Society of Petroleum Geologists, Calgary, Alberta, 1964.
(154) Kartsev, A. A., "Geochemical Transformation of Petroleum," in "Advances in Organic Geochemistry," pp. 11–14, Macmillan, New York, 1964.
(155) Philippi, G. T., "On the Depth, Time and Mechanism of Petroleum Generation," *Geochim. Cosmochim. Acta* (1965) **29** (9), 1021–1049.
(156) Philippi, G. T., "Essentials of the Petroleum Formation Process Are Organic Source Material and a Subsurface Temperature Controlled Chemical Reaction Mechanism," in "Advances in Organic Geochemistry," pp. 25–46, Pergamon, Oxford, 1969.
(157) Welte, D. H., "Relation Between Petroleum and Source Rock," *Am. Assoc. Petrol. Geol. Bull.* (1965) **49** (12), 2246–2268.
(158) Welte, D. H., "Correlation Problems Among Crude Oils," in "Advances in Organic Geochemistry," pp. 111–127, Pergamon, Oxford, 1970.
(159) Tissot, B., "Geochemical Problems in Petroleum Generation and Migration," *Rev. Inst. Franc. Petrole* (1966) **21** (11), 1621–1671.
(160) Tissot, B., Premieres Données sur les Mécanismes et la Cinétique de la Formation du Pétrole dan les Sédiments: Simulation d'un Schéma Réactionnel sur Ordinateur," *Rev. Inst. Franc. Petrole* (1969) **24** (4), 470–501.
(161) Mast, R. F., Shimp, N. F., Witherspoon, P. A., "Geochemical Trends in Chesterian (Upper Mississippian) Waltersburg Crudes of the Illinois Basin," *Illinois State Geol. Survey Circ.* (1968) **421.**
(162) Radchenko, O. A., "Geochemical Regularities in the Distribution of the Oil-Bearing Regions of the World," translation by Israel Program for Scientific Translations, Jerusalem, Israel, 1968.
(163) Byramjee, R. S., Bestougeff, M. A., "Etude sur les Transformations Physiques et Chimiques des Pétroles en Liaison Avec les Conditions Géologiques," in "Advances in Organic Geochemistry," pp. 129–151, Pergamon, Oxford, 1970.
(164) Hitchon, B., "Rock Volume and Pore Volume Data for Plains Region of Western Canada Sedimentary Basin between Latitudes 49° and 60°N.," *Am. Assoc. Petrol. Geol. Bull.* (1968) **52** (12), 2318–2323.
(165) Jeffrey, L. M., Hood, D. W., "Organic Matter in Sea Water; An Evaluation of Various Methods for Isolation," *J. Marine Res.* (1958) **17,** 274–271.
(166) Tatsumoto, M., Williams, W. T., Prescott, J. M., Hood, D. W., "Amino Acids in Samples of Surface Sea Water," *J. Marine Res.* (1961) **19** (2), 89–95.
(167) Williams, P. M., "Organic Acids in Pacific Ocean Waters," *Nature* (1961) **189** (4760), 219–220.

(168) Park, K., Williams, W. T., Prescott, J. M., Hood, D. W., "Amino Acids in Deep-Sea Water," *Science* (1962) **138** (3539), 531–532.
(169) Hood, D. W., "Chemical Oceanography," *Oceanogr. Marine Biol. Ann. Rev.* (1963) **1**, 129–155.
(170) Degens, E. T., "On Biochemical Processes in the Early Stage of Diagenesis," *Proc. Intern. Sedimentol. Congr., 6th* (1964), 83–92.
(171) Chau, Y. K., Riley, J. P., "The Determination of Amino-Acids in Sea Water," *Deep-sea Res.* (1966) **13** (6), 1115–1124.
(172) Bader, R. G., Hood, D. W., Smith, J. B., "Recovery of Dissolved Organic Matter in Sea-Water and Organic Sorption by Particulate Material," *Geochim. Cosmochim. Acta* (1960) **19** (4), 236–243.
(173) Peake, E., Hodgson, G. W., "Alkanes in Aqueous Systems. I. Exploratory Investigations on the Accommodation of C_{20}–C_{33} *n*-Alkanes in Distilled Water and Occurrence in Natural Water Systems," *Am. Oil Chemists Soc. J.* (1966) **43** (3), 215–222.
(174) Davis, J. B., Yarbrough, H. R., "Geochemical Exploration," U. S. Patent **3,457,044** (1969).
(175) McAuliffe, C., "Determination of Dissolved Hydrocarbons in Subsurface Brines," *Chem. Geol.* (1969) **4** (1/2), 225–233.
(176) Bykova, E. L., Nikitina, I. B., "Water-Soluble Organic Matter in Groundwater and Surface Water of South Yakutia," *Geokhimiya* (1964) **12,** 1299–1304.
(177) Davis, J. B., "Distribution of Naphthenic Acids in An Oil-Bearing Aquifer," *Chem. Geol.* (1969) **5** (2), 89–97.
(178) Zarrella, W. M., Mousseau, R. J., Coggeshall, N. D. *et al.*, "Analysis and Interpretation of Hydrocarbons in Sub-Surface Brines," Abstracts of Papers, Meeting, ACS, p. 50.
(179) Stadnik, E. V., "Use of Benzene in Ground Waters of the Lower Volga Region As a Criterion for Evaluation of Oil and Gas Prospects," *Geol. Neft. Gaz.* (1966) **4,** 43–46.
(180) Kartsev, A. A., Dudova, M. Ya., Diterikhs, O. D., "Benzene Homologs in Underground Waters and Their Relation To Oil," *Geol. Neft. Gaz.* (1969) **7**, 41–45.
(181) Degens, E. T., Hunt, J. M., Reuter, J. H., Reed, W. E., "Data on the Distribution of Amino Acids and Oxygen Isotopes in Petroleum Brine Waters of Various Geologic Ages," *Sedimentology* (1964) **3,** 199–225.
(182) Nuriev, A. N., Efendiev, G. K., "Composition of Organic Compounds Dissolved in Interstitial Waters of Oil Fields," *Dokl. Akad. Nauk Azerbaidzh. SSR* (1966) **22** (8), 29–31.
(183) Cooper, J. E., Kvenvolden, K. A., "Method for Prospecting for Petroleum," U. S. Patent **3,305,317** (1967).
(184) Baker, E. G., "Oil Migration in Aqueous Solution—A Study of the Solubility of n-Octadecane," *Am. Chem. Soc., Div. Petrol. Chem., Preprints* (1956) 5–17.
(185) Baker, E. G., "Crude Oil Composition and Hydrocarbon Solubility," *Am. Chem. Soc., Div. Petrol. Chem., Preprints* (1958) 61–68.
(186) Baker, E. G., "Origin and Migration of Oils," *Science* (1959) **129** (3353), 871–874.
(187) Baker, E. G., "A Hypothesis Concerning the Accumulation of Sediment Hydrocarbons to Form Crude Oil," *Geochim. Cosmochim. Acta* (1960) **19** (4), 309–317.
(188) Baker, E. G., "Distribution of Hydrocarbons in Petroleum," *Am. Assoc. Petrol. Geol. Bull.* (1962) **46** (1), 76–84.
(189) Baker, E. G., "A Geochemical Evaluation of Petroleum Migration and Accumulation," in "Fundamental Aspects of Petroleum Geochemistry," pp. 299–329, Elsevier, Amsterdam, 1967.

(190) Kennedy, W. A., "Solubilization of Hydrocarbons As a Process of Formation of Petroleum Deposits," Ph.D. Thesis (1964), Texas University.
(191) Peake, E., Hodgson, G. W., "Alkanes in Aqueous Systems. II. The Accommodation of C_{12}–C_{36} *n*-Alkanes in Distilled Water," *Am. Oil Chemists Soc. J.* (1967) **44** (12), 696–702.
(192) Hubbert, M. K., "The Theory of Ground-Water Motion," *J. Geol.* (1940) **48** (8), 785–944.
(193) Meyboom, P., "Current Trends in Hydrogeology," *Earth-Sci. Rev.* (1966) **2**, 345–364.
(194) Toth, J., "A Theory of Groundwater Motion in Small Drainage Basins in Central Alberta, Canada," *J. Geophys. Res.* (1962) **67** (11), 4375–4387.
(195) Toth, J., "A Theoretical Analysis of Groundwater Flow in Small Drainage Basins," *J. Geophys. Res.* (1963) **68** (16), 4795–4812.
(196) Freeze, R. A., Witherspoon, P. A., "Theoretical Analysis of Regional Groundwater Flow. 1. Analytical and Numerical Solutions to the Mathematical Model," *Water Resources Res.* (1966) **2** (4), 641–656.
(197) Freeze, R. A., Witherspoon, P. A., "Theoretical Analysis of Regional Groundwater Flow. 2. Effect of Water-Table Configuration and Subsurface Permeability Variation," *Water Resources Res.* (1967) **3** (2), 623–634.
(198) Freeze, R. A., Witherspoon, P. A., "Theoretical Analysis of Regional Groundwater Flow. 3. Quantitative Interpretations," *Water Resources Res.* (1968) **4** (3), 581–590.
(199) Freeze, R. A., "Theoretical Analysis of Regional Groundwater Flow," Ph.D. Thesis University of California, Berkeley, 1966.
(200) Freeze, R. A., "Theoretical Analysis of Regional Groundwater Flow," *Can. Inland Waters Branch, Sci. Ser.* (1969) **3**.
(201) Hitchon, B., "Fluid Flow in the Western Canada Sedimentary Basin. 1. Effect of Topography," *Water Resources Res.* (1969) **5** (1), 186–195.
(202) Hitchon, B., "Fluid Flow in the Western Canada Sedimentary Basin. 2. Effect of Geology," *Water Resources Res.* (1969) **5** (2), 460–469.
(203) Hitchon, B., Hays, J., "Hydrodynamics and Hydrocarbon Occurrences, Surat Basin, Queensland, Australia," *Water Resources Res.*, in press.
(204) Hitchon, B., Friedman, I., "Geochemistry and Origin of Formation Waters in the Western Canada Sedimentary Basin—1. Stable Isotopes of Hydrogen and Oxygen," *Geochim. Cosmochim. Acta* (1969) **33** (11), 1321–1349.
(205) Van Everdingen, R. O., "Studies of Formation Waters in Western Canada: Geochemistry and Hydrodynamics," *Can. J. Earth Sciences* (1968) **5**, 523–543.
(206) Stallman, R. W., "Notes on the Use of Temperature Data for Computing Groundwater Velocity," *Soc. Hydrotech. France, Assemblee Hydraulique, 6th* (1960) **3**, 1–7.
(207) Hitchon, B., "Geochemical Studies of Natural Gas. Part I. Hydrocarbons in Western Canadian Natural Gases," *J. Can. Petrol. Tech.* (1963) **2** (2), 60–76.
(208) Hitchon, B., "Geochemical Studies of Natural Gas. Part II. Acid Gases in Western Canadian Natural Gases," *J. Can. Petrol. Tech.* (1963) **2** (3), 100–116.
(209) Hitchon, B., "Geochemical Studies of Natural Gas. Part III. Inert Gases in Western Canadian Natural Gases," *J. Can. Petrol. Tech.* (1963) **2** (4), 165–174.
(210) Hitchon, B., "Geochemistry of Natural Gas in Western Canada," in "Natural Gases of North America," *Am. Assoc. Petrol. Geol., Tulsa, Okla., Mem.* (1968) **9, 2,** 1995–2025.

(211) Mangelsdorf, P. C., Manheim, F. T., Gieskes, J. M. T. M., "Role of Gravity, Temperature Gradients and Ion-Exchange Media in Formation of Fossil Brines," *Am. Assoc. Petrol. Geol. Bull.* (1970) **54** (4), 617–626.
(212) Watts, H., "The Possible Role of Adsorption and Diffusion in the Accumulation of Crude Petroleum Deposits: A Hypothesis," *Geochim. Cosmochim. Acta* (1963) **27** (8), 925–928.
(213) DeSitter, L. U., "Diagenesis of Oil-Field Brines," *Am. Assoc. Petrol. Geol. Bull.* (1947) **31** (11), 2030–2040.
(214) Wyllie, M. R. J., "Some Electrochemical Properties of Shales," *Science* (1948) **108** (2816), 684–685.
(215) Berry, F. A. F., "Hydrodynamics and Geochemistry of the Jurassic and Cretaceous Systems in the San Juan Basin, Northwestern New Mexico and Southwestern Colorado," Ph.D. Thesis (1958), Stanford University.
(216) Berry, F. A. F., "Relative Factors Influencing Membrane Filtration Effects in Geologic Environments," *Chem. Geol.* (1969) **4** (1/2), 295–301.
(217) Berry, F. A. F., Hanshaw, B. B., "Geologic Field Evidence Suggesting Membrane Properties of Shales," *Intern. Geol. Congr., 21st, Copenhagen* (1960), Abstracts, 209.
(218) McKelvey, J. G., Milne, I. H., "The Flow of Salt Solutions Through Compacted Clays," in "Clays and Clay Minerals," Vol. 9, pp. 248–259, Pergamon, Oxford, 1960.
(219) Hill, G. A., Colburn, W. A., Knight, J. W., "Reducing Oil-Finding Costs by Use of Hydrodynamic Evaluations," in "Economics of Petroleum Exploration, Development, and Property Evaluation," *Proc. 1961 Inst. Intern. Oil Gas Educ. Center,* Prentice-Hall, Englewood, 38–69.
(220) Hanshaw, B. B., "Membrane Properties of Compacted Clays," Ph.D. Thesis (1962), Harvard University.
(221) Hanshaw, B. B., "Cation-Exchange Constants for Clays from Electrochemical Measurements," in "Clays and Clay Minerals," Vol. 12, pp. 397–421, Macmillan, New York, 1964.
(222) Milne, I. H., McKelvey, J. G., Trump, R. P., "Permeability and Salt-Filtering Properties of Compacted Clay," in "Clays and Clay Minerals," Vol. 11, pp. 250–251, Pergamon, Oxford, 1962.
(223) Milne, I. H., McKelvey, J. G., Trump, R. P., "Semi-Permeability of Bentonite Membranes To Brines," *Am. Assoc. Petrol. Geol. Bull.* (1964) **48** (1), 103–105.
(224) Bredehoeft, J. D., Blyth, C. R., White, W. A., Maxey, G. B., "Possible Mechanism For Concentration of Brines in Subsurface Formations," *Am. Assoc. Petrol. Geol. Bull.* (1963) **47** (2), 257–269.
(225) Graf, D. L., Friedman, I., Meents, W. F., "The Origin of Saline Formation Waters. II: Isotopic Fractionation by Shale Micropore Systems," *Illinois State Geol. Survey, Circ.* (1965) **393.**
(226) Hanshaw, B. B., Zen, E-an, "Osmotic Equilibrium and Overthrust Faulting," *Geol. Soc. Am. Bull.* (1965) **76** (12), 1379–1386.
(227) White, D. E., "Saline Waters of Sedimentary Rocks," in "Fluids in Subsurface Environments," *Am. Assoc. Petrol. Geol., Tulsa, Okla. Mem.* (1965) **4,** 342–366.
(228) Young, A., Low, P. F., "Osmosis in Argillaceous Rocks," *Am. Assoc. Petrol. Geol. Bull.* (1965) **49** (7), 1004–1007.
(229) Hanshaw, B. B., Hill, G. A., "Geochemistry and Hydrodynamics of the Paradox Basin Region, Utah, Colorado and New Mexico," *Chem. Geol.* (1969) **4** (1/2), 263–294.

(230) Amstutz, G. C., Bubenicek, L., "Diagenesis in Sedimentary Mineral Deposits," in "Diagenesis in Sediments," pp. 417–475, Elsevier, Amsterdam, 1967.
(231) Noble, E. A., "Formation of Ore Deposits by Water of Compaction," *Econ. Geol.* (1963) **58** (7), 1145–1156.
(232) Brown, J. S., Ed., "Genesis of Stratiform Lead-Zinc-Barite-Fluorite Deposits (Mississippi Valley Type Deposits)—A Symposium," *Econ. Geol.* (1967) Monograph **3.**
(233) Heyl, A. V., "Some Aspects of Genesis of Zinc-Lead-Barite-Fluorite Deposits in the Mississippi Valley, U.S.A.," *Trans. Sect. B., Inst. Mining Metallurgy* (1969) **78,** B148–B160.
(234) Jackson, S. A., Beales, F. W., "An Aspect of Sedimentary Basin Evolution: The Concentration of Mississippi Valley-Type Ores During Late Stages of Diagenesis," *Bull. Canadian Petrol. Geol.* (1967) **15** (4), 383–433.
(235) Cumming, G. L., Robertson, D. K., "Isotopic Composition of Lead From the Pine Point Deposit," *Econ. Geol.* (1969) **64** (7), 731–732.
(236) Fritz, P., "The Oxygen and Carbon Isotopic Composition of Carbonates From the Pine Point Lead-Zinc Ore Deposits," *Econ. Geol.* (1969) **64** (7), 733–742.
(237) Jackson, S. A., Folinsbee, R. E., "The Pine Point Lead-Zinc Deposits, N.W.T., Canada. Introduction and Paleoecology of the Presqu'ile Reef," *Econ. Geol.* (1969) **64** (7), 711–717.
(238) Sasaki, A., Krouse, H. R., "Sulfur Isotopes and the Pine Point Lead-Zinc Mineralization," *Econ. Geol.* (1969) **64** (7), 718–730.
(239) Billings, G. K., Kesler, S. E., Jackson, S. A., "Relation of Zinc-Rich Formation Waters, Northern Alberta, To the Pine Point Ore Deposit," *Econ. Geol.* (1969) **64** (4), 385–391.
(240) Saxby, J. D., "Metal-Organic Chemistry of the Geochemical Cycle," *Rev. Pure Appl. Chem.* (1969) **19,** 131–150.
(241) Erdman, J. G., Harju, P. H., "Capacity of Petroleum Asphaltenes to Complex Heavy Metals," *J. Chem. Eng. Data* (1963) **8** (2), 252–258.
(242) Ahrens, L. H., "Ionization Potentials and Metal-Amino Acid Complex Formation in the Sedimentary Cycle," *Geochim. Cosmochim. Acta* (1966) **30** (11), 1111-1119.
(243) Evans, W. D., "The Organic Solubilization of Minerals in Sediments," in "Advances in Organic Geochemistry," pp. 263–270, Macmillan, New York, 1964.
(244) Jeffrey, L. M., Hood, D. W., "Chemistry of Organo-Silicate Complexes Isolated From Marine Sediments," *Abst., Geol. Soc. Am. Ann. Mtg., 1964, Miami Beach,* 100–101.
(245) Hood, D. W., "Organic-Inorganic Interactions in the Aquatic Environment," *Abst., Ann. Geol. Soc. Am. Assoc. Soc. Joint Mtg., 1967,* New Orleans, 102.
(246) Davis, D. R., "The Measurement and Evaluation of Certain Trace Metal Concentrations in the Nearshore Environment of the Northwest Gulf of Mexico and Galveston Bay," Ph.D. Thesis (1968), Texas A & M University.
(247) Ganeev, I. G., "On the Possible Transport of Matter in the Form of Complicated Complex Compounds," *Geokhimiya* (1962) **10,** 917–924.
(248) Rawson, J., "Solution of Manganese Dioxide by Tannic Acid," *U.S. Geol. Survey Prof. Paper* (1963) **475-C,** 218–219.
(249) Solomon, D. H., "Clay Minerals As Electron Acceptors and/or Electron Donors in Organic Reactions," in "Clays and Clay Minerals," Vol. 16, pp. 31–39, Pergamon, Oxford, 1968.
(250) Blumer, M., Mullin, M. M., Thomas, D. W., "Pristane in Zooplankton," *Science* (1963) **140** (3562), 974.

(251) Blumer, M., Mullin, M. M., Thomas, D. W., "Pristane in the Marine Environment," *Helgol. Wiss. Meeresunters.* (1964) **10** (1–4), 187–201.
(252) Blumer, M., Thomas, D. W., "Phytadienes in Zooplankton," *Science* (1965) **147** (3662), 1148–1149.
(253) Blumer, M., Thomas, D. W., " 'Zamene,' Isomeric C_{19} Monoolefins From Marine Zooplankton, Fishes, and Mammals," *Science* (1965) **148** (3668), 370–371.
(254) Blumer, M., Snyder, W. D., "Isoprenoid Hydrocarbons in Recent Sediments: Presence of Pristane and Probable Absence of Phytane," *Science* (1965) **150** (3703), 1588–1589.
(255) Eglinton, G., Scott, P. M., Belsky, T., Burlingame, A. L., Calvin, M., "Hydrocarbons of Biological Origin From a One-Billion-Year-Old Sediment," *Science* (1964) **145** (3629), 263–264.
(256) Eglinton, G., Scott, P. M., Belsky, T., Burlingame, A. L., Richter, W., Calvin, M., "Occurrence of Isoprenoid Alkanes in a Precambrian Sediment," in "Advances in Organic Geochemistry," pp. 41–74, Pergamon, Oxford, 1966.
(257) Meinschein, W. G., Barghoorn, E. S., Schopf, J. W., "Biological Remnants in a Precambrian Sediment," *Science* (1964) **145** (3629), 262–263.
(258) Meinschein, W. G., "Soudan Formation: Organic Extracts of Early Precambrian Rocks," *Science* (1965) **150** (3696), 601–605.
(259) Oró, J., Nooner, D. W., Zlatkis, A., Wikstrom, S. A., Barghoorn, E. S., "Hydrocarbons of Biological Origin in Sediments About Two Billion Years Old," *Science* (1965) **148** (3666), 77–79.
(260) McCarthy, E. D., Calvin, M., "The Isolation and Identification of the C_{17} Isoprenoid 2,6,10-Trimethyltetradecane from a Devonian Shale: The Role of Squalene as a Possible Precursor," *Tetrahedron* (1967) **23**, 2609–2619.
(261) Oró, J., Nooner, D. W., "Aliphatic Hydrocarbons in Pre-Cambrian Rocks," *Nature* (1967) **213** (5081), 1082–1085.
(262) Oró, J., Nooner, D. W., "Aliphatic Hydrocarbons from the Pre-Cambrian of North America and South Africa," in "Advances in Organic Geochemistry," pp. 493–506, Pergamon, Oxford, 1970.
(263) Modzeleski, V. E., MacLeod, W. D., Nagy, B., "A Combined Gas Chromatographic-Mass Spectrometric Method for Identifying *n*- and Branched-Chain Alkanes in Sedimentary Rocks," *Anal. Chem.* (1968) **40** (6), 987–989.
(264) Reed, W. E., Gilbert, C. M., "Organic Geochemistry of Some Late Pleistocene Sediments, Mono Basin, California," *Ann. Mtg. Geol. Soc. Am., 65th, Cordilleran Sect., Paleontol. Soc. Pacific Coast Sect. Mtg., 1969, Eugene, Ore.*, Program, 56–57.
(265) Cummins, J. J., Robinson, W. E., "Normal and Isoprenoid Hydrocarbons Isolated From Oil-Shale Bitumen," *J. Chem. Eng. Data* (1964) **9** (2), 304–307.
(266) Robinson, W. E., Cummins, J. J., Dinneen, G. U., "Changes in Green River Oil-Shale Paraffins With Depth," *Geochim. Cosmochim. Acta* (1965) **29** (4), 249–258.
(267) Douglas, A. G., Eglinton, G., Maxwell, J. R., "The Organic Geochemistry of Certain Samples From the Scottish Carboniferous Formation," *Geochim. Cosmochim. Acta* (1969) **33** (5), 579–590.
(268) Bendoraitis, J. G., Brown, B. L., Hepner, L. S., "Isoprenoid Hydrocarbons in Petroleum," *Anal. Chem.* (1963) **34**, 49–53.

(269) Bendoraitis, J. G., Brown, B. L., Hepner, L. S., "Isolation and Identification of Isoprenoids in Petroleum," *World Petrol. Congr., 6th* (1963) Sect. **5**, 158–159.

(270) Belsky, T., Johns, R. B., McCarthy, E. D., Burlingame, A. L., Richter, W., Calvin, M., "Evidence of Life Processes in a Sediment Two and a Half Billion Years Old," *Nature* (1965) **206** (4983), 446–447.

(271) Johns, R. B., Belsky, T., McCarthy, E. D., Burlingame, A. L., Haug, P., Schnoes, H. K., Richter, W., Calvin, M., "The Organic Geochemistry of Ancient Sediments—Part II," *Geochim. Cosmochim. Acta* (1966) **30** (12), 1191–1222.

(272) Van Hoeven, W., Haug, P., Burlingame, A. L., Calvin, M., "Hydrocarbons from Australian Oil, Two Hundred Million Years Old," *Nature* (1966) **211** (5056), 1361–1365.

(273) Goehring, K. E. H., Schenck, P. A., Engelhardt, E. D., "A New Series of Isoprenoid Isoalkanes in Crude Oils and Cretaceous Bituminous Shales," *Nature* (1967) **215** (5100), 503–505.

(274) Han, J., Calvin, M., "Occurrence of C_{22}–C_{25} Isoprenoids in Bell Creek Crude Oil," *Geochim. Cosmochim. Acta* (1969) **33** (6), 733–742.

(275) Blumer, M., Cooper, W. J., "Isoprenoid Acids in Recent Sediments," *Science* (1967) **158** (3807), 1463–1464.

(276) Eglinton, G., Douglas, A. G., Maxwell, J. R., Ramsay, J. N., Stallberg-Stenhagen, S., "Occurrence of Isoprenoid Fatty Acids in the Green River Shale," *Science* (1966) **153** (3731), 1133–1134.

(277) Haug, P., Schnoes, H. K., Burlingame, A. L., "Isoprenoid and Dicarboxylic Acids Isolated From Colorado Green River Shale (Eocene)," *Science* (1967) **158** (3802), 772–773.

(278) Douglas, A. G., Douraghi-Zadeh, K., Eglinton, G., Maxwell, J. R., Ramsay, J. N., "Fatty Acids in Sediments, Including the Green River Shale (Eocene) and Scottish Torbanite (Carboniferous)," in "Advances in Organic Geochemistry," pp. 315–334, Pergamon, Oxford, 1970.

(279) Cason, J., Graham, D. W., "Isolation of Isoprenoid Acids From a California Petroleum," *Tetrahedron* (1965) **21**, 471–483.

(280) Breger, I. A., "Geochemistry of Lipids," *Am. Oil Chem. Soc.* (1966) **43** (4), 196–202.

(281) Schwendinger, R. B., Erdman, J. G., "Sterols in Recent Aquatic Sediments," *Science* (1964) **144** (3626), 1575–1576.

(282) Smith, L. L., "Sterols in Aquatic Sediments," *Am. Chem. Soc., Petrol. Res. Fund, Ann. Rept., 11th* (1966), 180.

(283) Attaway, D. H., Parker, P. L., "Sterols in Recent Marine Sediments," *Abstract, Geol. Soc. Am. Ann. Mtg., Program* (1969) Pt. 7, 4–5.

(284) Stevenson, F. J., "Lipids in Soil," *Am. Oil Chem. Soc.* (1966) **43** (4), 203–210.

(285) Burlingame, A. L., Haug, P., Belsky, T., Calvin, M., "Occurrence of Biogenic Steranes and Pentacyclic Triterpanes in An Eocene Shale (52 Million Years) and in An Early Precambrian Shale (2.7 Billion Years); A Preliminary Report," *Proc. Natl. Acad. Sci.* (1965) **54,** 1406–1412.

(286) Hills, I. R., Whitehead, E. V., Anders, D. E., Cummins, J. J., Robinson, W. E., "An Optically Active Triterpane, Gammacerane in Green River, Colorado, Oil Shale Bitumen," *Chem. Commun.* (1966), 752.

(287) Anderson, P. C., Gardner, P. M., Whitehead, E. V., Anders, D. E., Robinson, W. E., "The Isolation of Steranes From Green River Oil Shale," *Geochim. Cosmochim. Acta* (1969) **33** (10), 1304–1307.

(288) Murphy, M. T. J., McCormick, A., Eglinton, G., "Perhydro-β-Carotene in the Green River Shale," *Science* (1967) **157** (3792), 1040–1042.

(289) Hills, I. R., Smith, G. W., Whitehead, E. V., "Optically Active Spirotriterpane in Petroleum Distillates," *Nature* (1968) **219** (5151), 243–246.

(290) Bergmann, W., "Geochemistry of Lipids," in "Organic Geochemistry," pp. 503–542, Pergamon, Oxford, 1963.
(291) Hoering, T. C., "Optically Active Steranes in a Miocene Petroleum," *Carnegie Inst. Washington Year Book 68, Papers Geophys. Lab.* (1970) **1560,** 303–307.
(292) Hodgson, G. W., Hitchon, B., Elofson, R. M., Baker, B. L., Peake, E., "Petroleum Pigments From Recent Fresh-Water Sediments," *Geochim. Cosmochim. Acta* (1960) **19** (4), 272–288.
(293) Thomas, D. W., Blumer, M., "Porphyrin Pigments of a Triassic Sediment," *Geochim. Cosmochim. Acta* (1964) **28** (7), 1147–1154.
(294) Barghoorn, E. S., Meinschein, W. G., Schopf, J. W., "Palaeobiology of a Precambrian Shale," *Science* (1965) **148** (3669), 461–472.
(295) Eggelpoel, A. van, "Comparison Entre les Porphyrines Extraites de la Roche-Reservoir (Argiles Silicifiees) d'Ozouri (Gabon) et Celles de l'Huite," in "Advances in Organic Geochemistry," pp. 227–242, Pergamon, Oxford, 1966.
(296) Blumer, M., Snyder, W. D., "Porphyrins of High Molecular Weight in a Triassic Oil Shale: Evidence by Gel Permeation Chromatography," *Chem. Geol.* (1967) **2** (1), 35–45.
(297) Kvenvolden, K. A., Hodgson, G. W., "Evidence for Porphyrins in Early Precambrian Swaziland System Sediments," *Geochim. Cosmochim. Acta* (1969) **33** (10), 1195–1202.
(298) Gransch, J. A., Eisma, E., "Geochemical Aspects of the Occurrence of Porphyrins in West Venezuelan Mineral Oils and Rocks," in "Advances in Organic Geochemistry," pp. 69–86, Pergamon, Oxford, 1970.
(299) Baker, E. W., "Mass Spectrometric Characterization of Petroporphyrins," *J. Am. Chem. Soc.* (1966) **88** (10), 2311–2315.
(300) Millson, M. F., Montgomery, D. S., Brown, S. R., "An Investigation of the Vanadyl Porphyrin Complexes of the Athabasca Oil Sands," *Geochim. Cosmochim. Acta* (1966) **30** (2), 207–221.
(301) Morandi, J. R., Jensen, H. B., "Comparison of Porphyrins from Shale Oil, Oil Shale and Petroleum by Absorption and Mass Spectroscopy," *J. Chem. Eng. Data* (1966) **11** (1), 81–88.
(302) Hodgson, G. W., Flores, J., Baker, B. L., "The Origin of Petroleum Porphyrins: Preliminary Evidence for Protein Fragments Associated with Porphyrins," *Geochim. Cosmochim. Acta* (1969) **33** (4), 532–535.
(303) McCamis, J. G., Griffith, L. S., "Middle Devonian Facies Relationships, Zama Area, Alberta," *Bull. Canadian Petrol. Geol.* (1967) **15** (4), 434–467.

RECEIVED August 31, 1970. Research Council of Canada Contribution No. 513.

3

Generation of Light Hydrocarbons and Establishment of Normal Paraffin Preferences in Crude Oils

M. A. ROGERS

Imperial Oil Limited, 500 Sixth Avenue S.W., Calgary, Alberta, Canada

C. B. KOONS

Esso Production Research Company, P.O. Box 2189, Houston, Tex. 77001

Oil forms by incorporation of heavy (C_{15+}) hydrocarbons from pre-existing biologic material and by generation of both light (C_4–C_7) and heavy (C_{15+}) hydrocarbons from bulk organic matter. Light hydrocarbons are absent in Recent sediments; they are only formed over finite intervals of time and depth. Three stages are observed in young sediments: (1) no light hydrocarbons, only bacterially produced gases; (2) no light hydrocarbons but heptane, the smallest naturally occurring hydrocarbon, preservation and incorporation of hydrocarbons; (3) some light hydrocarbons, but these are much attenuated with respect to gases and heptane; onset of generation. Normal alkanes generated from marine organic matter do not have the odd-carbon preference of terrestrial organic matter. This unique feature arises from the strong bacterial impress during organic matter transformations.

Crude oil contains heavy hydrocarbons incorporated from pre-existing biological material and both light and heavy hydrocarbons formed from bulk organic material. Because light hydrocarbons are absent in Recent sediments, they must be formed over finite intervals of time and depth. Heavy hydrocarbons have been transformed, but many still reflect the precursor organic molecules.

For convenience we define heavy hydrocarbons as those with more than 15 carbon atoms per molecule (C_{15+}) and light hydrocarbons as those with four to seven carbon atoms per molecule (C_4–C_7).

The presence of hydrocarbons in Recent marine sediments was demonstrated by Smith (*1*), and at that time it was generally acknowledged that petroleum originated from biogenic sources. It was also assumed that the process(es) of formation could be traced through young sediments. Dunton and Hunt (*2*), however, noted that Recent sediments did not contain one of the most abundant fractions in crude petroleum, light hydrocarbons between propane (C_3) and *n*-heptane (C_7). The samples on which their studies were based were of necessity limited to those taken in shallow waters and nearshore sites, and these contained large amounts of terrestrial organic carbon.

From a recent geochemical investigation of sediments from the continental slope in the northern Gulf of Mexico we have been able to recognize and distinguish between "marine organic facies" and "terrestrial organic facies" in marine sediments on the basis of their C_{15+} *n*-paraffin content. We report here our attempts to codify these two organic realms and to explain their distinguishing characteristics.

Our study was based on samples from 11 (Figure 1) 1000-foot core holes drilled in the Gulf of Mexico by four oil companies: Humble, Chevron, Gulf, and Mobil. All cores are from the present continental slope within three morphological areas: the Upper Continental Slope off Texas and Louisiana, the Upper Continental Slope off west Florida, and the upper reaches of the Mississippi Cone—a mass of sediment derived from drainage of the Mississippi River which has locally buried the continental-slope morphology.

Initial geochemical analyses indicated that organic matter from these slope sediments differed from previously investigated organic matter (from nearshore Recent sediments, older sediments, and rocks) in three significant respects:

(1) Although light hydrocarbons are not present in the surficial sequences, hydrocarbons with molecular weights between those of butane and heptane are observed in deeper sequences. We are able to establish the generation sequence in these sediments.

(2) *n*-Paraffins from slope sediments which probably contain no terrestrial organic matter show no preference for higher concentrations of molecules with an odd number of carbon atoms (C_{23}, C_{25}, C_{29}, etc.). Conversely, the literature, as well as some of our data, show that in nearshore sediments odd-numbered paraffins predominate.

(3) Carbon–isotopic ($^{13}C/^{12}C$) ratios of the combustible organic matter fluctuate, but these variations can be correlated with warm (interglacial) and cold (glacial) periods of the Pleistocene (*3*).

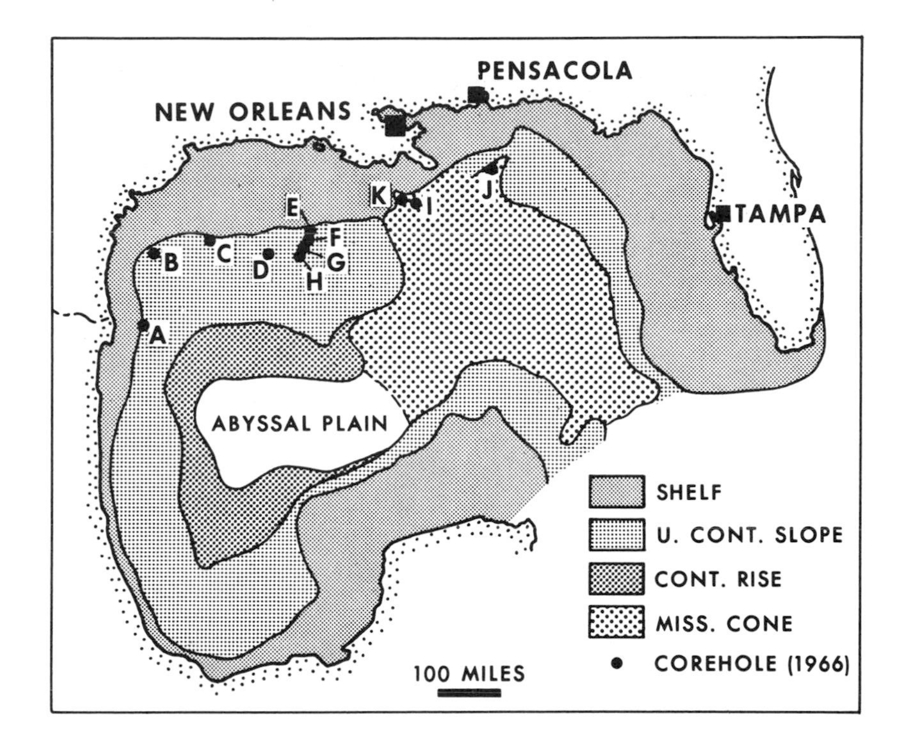

Figure 1. Map of sample locations

Light-Hydrocarbon Contents of Slope Sediments

Light hydrocarbons were extracted, isolated, and characterized chromatographically following techniques of Dunton and Hunt (2). Sediment studies require no special modifications other than more stringent precautions to avoid contamination because of the lower concentrations in sediments than rocks. Figures 2 and 3 are both based on analyses from the same technique, but in Figure 3 the chromatograms were produced as part of a routine, automatic operation. Extracts are encapsulated in indium and then introduced into the heated injection block; presentation is somewhat compressed, but results are comparable.

From our study we can characterize the marine organic realm on the basis of light-hydrocarbon generation sequences. The generation sequence for light hydrocarbons in young sediments is shown in Figure 2. The sequential arrangement of chromatograms A–C represents schematically our notion of what occurs with generation. The illustration is realistic because each chromatogram is from an actual sample. Unfortunately all changes cannot be illustrated with chromatograms from a single core hole; the sequence must be compiled from several overlapping

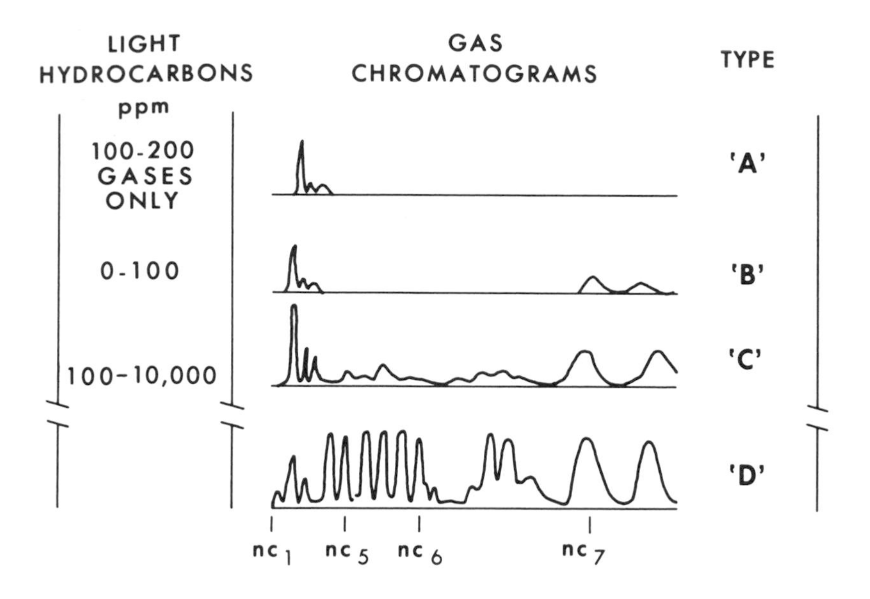

Figure 2. Generation sequence for light hydrocarbons in young sediments

holes. Each chromatogram, however, is representative of many samples from the Gulf of Mexico continental slope, although some are more common than others.

Chromatogram A is most common in the uppermost sediments, but some entire cores show nothing but this pattern of only the light gases methane, ethane, and propane. The presence of gases in young sediments is not remarkable: methane is ubiquitous. Its origin, in almost every terrestrial and marine environment, is demonstratively microbial, and ethane and propane may be formed in the same way. However, since it has never been demonstrated that either ethane or propane can be formed from microbes, it is more likely that ethylene and propylene (both of which are common products of plant growth) are in some manner reduced after adsorption in sediments. Nevertheless, it is unlikely that detection of these gases is significant in tracing the generation of petroleum hydrocarbons.

Chromatogram B is less common than A but is still not rare. Although the gases are present, no other light hydrocarbons are detected smaller than *n*-heptane (n-C_7). Often other hydrocarbons of higher molecular weight (such as methylcyclohexane) are detected beyond *n*-heptane. The *n*-heptane (n-C_7) hydrocarbon is the smallest that occurs naturally in living organisms. Hydrocarbons of higher molecular weight are also known to be present in cells. Chromatogram B represents, then, merely the preservation of hydrocarbons already present in organic matter incorporated at the time of sedimentation.

The third type of chromatogram, C, is relatively rare. Presentation is similar to that for A and B; however, there are additional peaks between butane (C_4) and heptane (C_7). All these peaks are much smaller than heptane or methylcyclohexane. Pattern C represents the onset of generation of light hydrocarbons. Heptane and methylcyclohexane are still relatively more abundant because generation has not progressed very far. Most heptane and methylcyclohexane here are probably preserved hydrocarbon, whereas all of the C_4 through C_6 compounds have been generated. Sediments showing chromatograms such as C represent the beginning of generation—*i.e.*, maturation. Only when we see this pattern can we say unequivocally that maturation has begun.

The petroleum generation sequence can also be followed approximately by increases in the total abundance of the light hydrocarbons. Type A has no light hydrocarbons and between 100 and 200 ppb of the gases. Type B, with about equivalent amounts of gases, has 49 ppb hep-

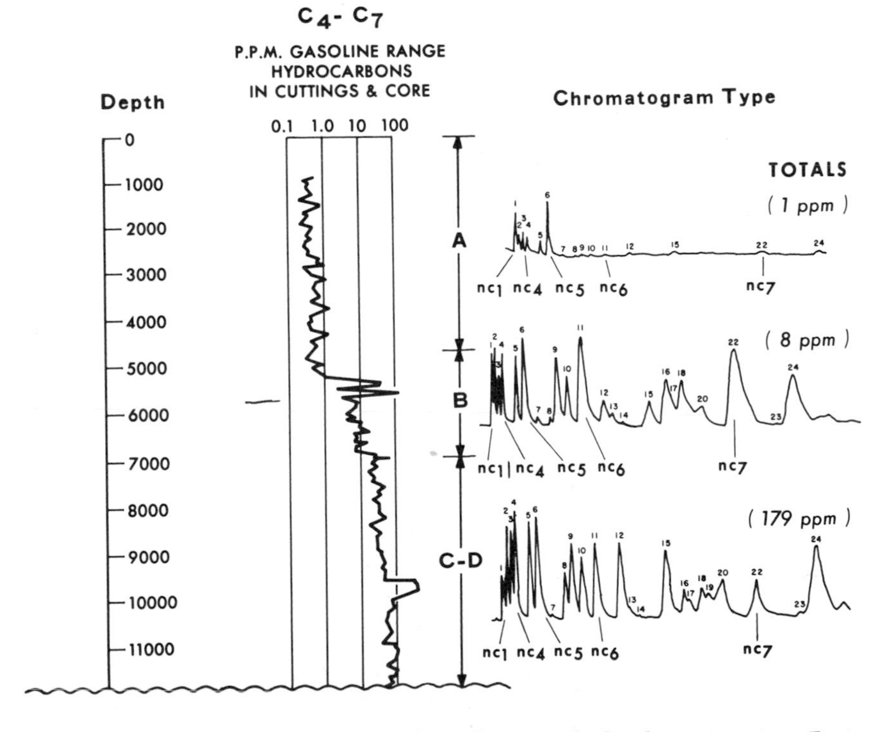

Figure 3. Generation of light hydrocarbons with depth in Beaufort Basin, N.W.T. Canada (based on a representative gas-cuttings log)

tane, methylcyclohexane, and some higher molecular weight hydrocarbons. The C has 300–10,000 ppb light hydrocarbons. For comparison, the chromatogram of a typical crude oil is included in Figure 2.

This increase by generation in total C_4–C_7 is apparent in basins containing young sediments. Consider Figure 3, which shows the progressive increase with depth for the Beaufort Basin of northern Canada. Although we have here interpreted generation stages on the basis of total C_4–C_7, detailed chromatography of samples from any one well leads to the same conclusions as those derived from the Gulf of Mexico study.

Normal Paraffin Content of Slope Sediments

Heavy hydrocarbons were obtained by solvent extraction (*4*) of sediments, deasphalting with pentane, and separation by liquid chromatography (*5*) into saturate, aromatic, NSO-eluted, and asphaltene fractions. Saturate fractions were analyzed by gas–liquid chromatography (*6*); on these chromatograms (Figures 4 and 6) *n*-paraffins "stand up" as peaks above the naphthenic background.

Previous studies on heavy hydrocarbon content of young sediments showed that they contained all the *n*-paraffin series, but that those containing paraffins with odd numbers of carbon atoms were several times as abundant as those containing even numbers of carbon atoms. Bray and Evans (*7*) defined a ratio of odd to even paraffins as a carbon preference index (CPI). In Recent sediments (mostly from nearshore environments) the CPI varied from 2.5 to 5.5. Crude oils and extracts from ancient shales exhibited little odd-carbon preference. The CPI for ancient shales varied from 0.9 to 2.3 and that for crude oils from 0.6 to 2.2

Some *n*-paraffin analyses on the soluble organic extract of the Gulf slope sediments are presented in Table I. It was totally unexpected that heavy hydrocarbon (C_{15+}) extracts from these sediments would have a CPI of approximately 1.0 (instead of >2.5 which is usually suggested for young sediments).

Figure 4 is a sequence of *n*-paraffin chromatograms on the saturate fraction of sediment extracts for core hole G. Note that although there is very little change in the CPI (it remains about 1.0), the hydrocarbon spectrum changes with depth as the *n*-paraffins between *n*-C_{15} and *n*-C_{18} appear. The predominant *n*-paraffin shifts to a lower molecular weight. By this latter criterion of Welte (*8*), the lowermost segments are more "mature" with essentially no change in CPI.

It has been postulated that odd-carbon preference is characteristic of bulk organic matter and that its disappearance results from thermal maturation processes as petroleum is being formed. However, as these

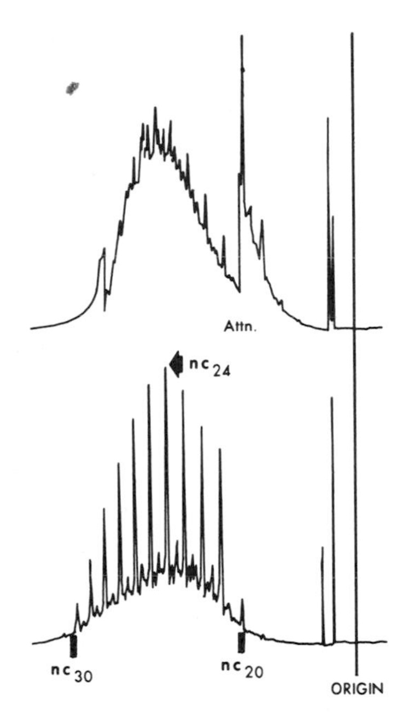

SEGMENT 7
NO C.P.I. CALCULATED
DEPTH (ft. below sea bottom)
317-332

SEGMENT 12
STANDARD C.P.I. = 1.04
DEPTH (ft. below sea bottom)
534-549

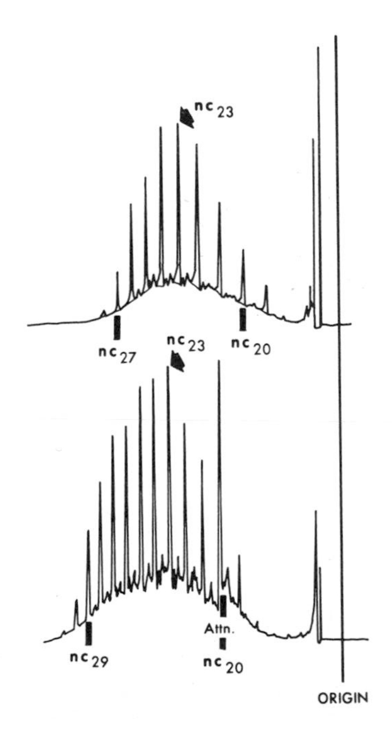

SEGMENT 18
SELECT. PEAK. C.P.I. = 0.93
DEPTH (ft. below sea bottom)
770-786

SEGMENT 24
STANDARD C.P.I. = 1.06
DEPTH (ft. below sea bottom)
967-999

Figure 4. Chromatograms for saturated hydrocarbon fractions showing predominance of n-*paraffins (sequence from core G)*

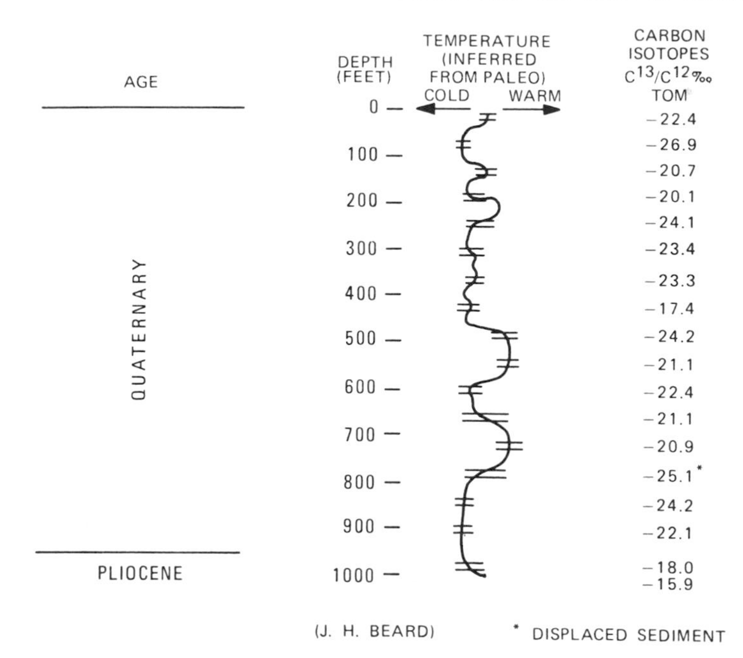

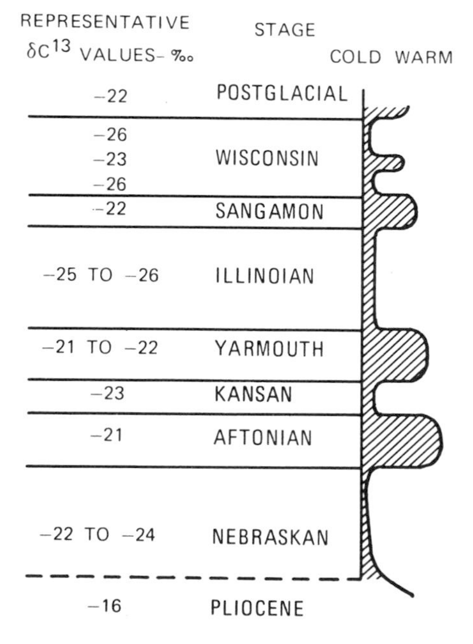

Figure 5. Carbon isotopic and paleotemperature variations. $^{13}C/^{12}C$ ratio data on core hole J (top) and for glacial and interglacial (bottom).

data show, such maturation processes cannot account for the low CPI seen in these young sediments.

Resolution of the anomalies between our analytical results and others led us to the postulate that these slope sediments contained organic matter which had not only been deposited in the marine organic realm but had also originated there. (Most previous analyses reported in the literature were on organic matter either from the terrestrial organic realm or from zones where there was admixing.)

Basic to further discussion of this postulate is some understanding of the contribution terrestrially derived organic matter actually makes to the total organic matter in marine sediments. If it should be a major one, many of our anomalies could result from the following process. During glacial periods sea level is lowered, and rivers discharge sediments and organic matter on the outer edge of the continental shelf. Terrestrially derived organic matter then might make a sizable contribution to marine sediments on the slope. Turbidity currents could carry some of the organic matter to the bottom of the slope and out onto the abyssal plain. Then, even in the deepest water sediments, we might observe organic matter from shore and nearshore environments. However, during interglacial periods, with high sea levels, our reasoning leads us to expect that at any one site substantially less terrestrially derived material must be incorporated in the sediments than in the enclosing glacials. Actually, neither *n*-paraffin preferences nor light hydrocarbon sequences show any change as we pass through glacial and interglacial sequences in deep water sediments.

Recently (*3*) we presented evidence which establishes the correlation between carbon isotope ratios and warm and cold periods within the Pleistocene. (Warm and cold periods are determined independently by paleontology.) Because the major difference between interglacial and glacial periods is in surface water temperature, we conclude that carbon isotopic fluctuations are caused primarily by variations in temperature of the photosynthetic zone where the organic matter in slope-sediment samples has been generated. The varying temperatures of this zone are reflected not only by the $^{13}C/^{12}C$ ratios of the total organic matter but also by the co-existing microorganisms (now fossil). Carbon isotopic values are characteristically glacial or interglacial, regardless of the amount of terrestrial contribution, closeness to shore, etc.

Figure 5 illustrates the dependence of the isotopic ratio on surface water temperature for core hole J.

The role of land masses in supplying marine basins with organic matter has been studied by other workers, particularly the Russians. Skopintsev (*9*) estimates that less than 1% of organic matter contained in basins and oceans is carried in by rivers. Even in inland basins such as

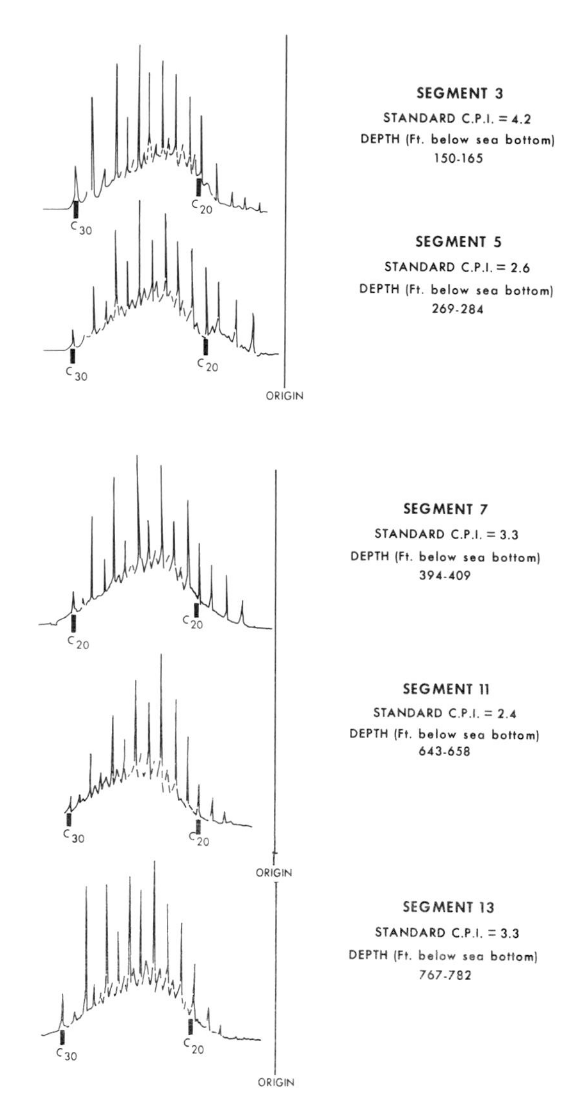

Figure 6. Chromatograms for saturated hydrocarbon fractions showing predominance of n-*paraffins (sequence from core K)*

the Sea of Azov and the Caspian Sea, according to Datsko (*10*), only about 3% of the organic matter is derived from river waters. Bordovskiy (*11*) states that mainland runoff is far more important in supplying rain basins with the nutrient salts essential to the development of phytoplankton than in supplying them with organic matter. We conclude that in the contribution of organic matter to marine basins land masses have played a negligible role, and assume that these comments apply equally to the deep water sequences of the Gulf of Mexico.

For comparison, a core (K) in the Mississippi Canyon was studied; this site on the upper continental slope has likely had the maximum contribution of terrestrial sediments. The *n*-paraffin chromatograms showed here, as expected, a marked odd-carbon preference (Figure 6).

Our explanation for the absence of odd-carbon preference in marine organic matter is based on the fact that this absence is also seen in extracts of marine organisms and of bacterial cells. Our evidence from slope sediments suggests that the strong bacterial impression on normal marine, deep water sedimentary organic matter produces both features.

Koons, Jamieson, and Ciereszko (*12*) recognized a CPI of approximately 1.0 from many marine organisms. The work of Davis (*13*) documented a CPI of 1.0 for hydrocarbons contained in bacterial cell extracts from the anaerobic sulfate-reducing bacterial *Desulfovibrio desulfuricans*. Davis extracted lipid material which constitutes about 5–9% of the cellular weight. About 25% of this lipid material is saturated hydrocarbons; the rest is primarily fatty acids, some of which are probably unsaturated. Among the hydrocarbons there was a prominent series of *n*-paraffins in the C_{25} to C_{35} range, which interestingly enough, showed no particular preference for odd- or even-numbered carbon chains—*i.e.*, the CPI was approximately 1.0. However, among the fatty acids the most abundant were those of shorter chain lengths—*e.g.*, the palmitic (C_{16}), stearic (C_{18}), and oleic ($C_{18:1}$). This disparity between the chain lengths of the *n*-paraffins and those of the straight chain fatty acids led Davis to conclude that there was no obvious precursor-to-product relationship between the fatty acids and the paraffins in the bacterial, cellular lipids (a mechanism for *n*-paraffin generation, with attendant diminution of the odd-carbon preference—at least in sediments—which had been previously

Table I. Selected *n*-Paraffin Analyses of Bitumen Extract

Core, id.	*Depth Below Sea Bottom, ft.*	*CPI*
A	760	1.39
B	675	0.94
B	710	1.01
C	530	1.06
D	10	0.90
E	650	1.08
E	810	1.10
F	740	1.23
G	360	0.91
H	400	0.85

Table II. Schematic for the Origin, Incorporation, and

Preferred Terminology	Wates Depth Environments
Terrestrial organic facies	Shallow Marine
	Near Shore
	Displaced sequences ≅ Turbidites
Marine organic facies	Deep Marine
	Offshore

postulated). However, in deeper water environments one might expect an extensive alteration of the existing organic matter *via* bacteria—*i.e.*, bacterial synthesis of hydrocarbons and other products. Being last in the food chain, bacteria ultimately become the only organisms capable of biosynthesis and thus would be expected to affect organic matter modifications as long as biospheric conditions persist. Theoretically, this effect could be expected to decrease with time and depth of sediments (to some limiting depth). As bacterial synthetic patterns gradually displace or are superimposed on the pattern of previous plant and animal biosynthesis, fatty acids in the n-$_{16}$ to n-C_{20} range become the source for the bacterial cellular hydrocarbons without marked CPI in the C_{25} to C_{35} range. These hydrocarbons are then liberated.

Summary

We have developed a schematic model for the origin, incorporation, and characterization of organic matter in marine sediments (Table II). We believe that truly marine organic matter originates by phytoplanktonic photosynthesis in the overlying water column and is then incorporated with the sediments being deposited locally. This organic matter is degraded in the uppermost layers of sediments, primarily by bacteria. These bacterial transformations produce hydrocarbons, particularly par-

Characterization of Organic Matter in Marine Sediments

Source of Organic Materials	*Characteristics of Organic Materials*
C_{15+}: land photosynthesis source predominates. C_{4-7}: generally not present, may be developed (generated) with depth (time).	Carbon isotopes generally more negative $\delta^{13}C$ -27 to -29 ‰
Products of higher plants and animals in high concentration, bacteria in "relatively" low concentration	n-Papaffins; marked odd CPI, complete range n-C_{15}–n-C_{32} Fatty acids; even CPI, complete range C_{16}–C_{32}
C_{15+}: marine phytoplantonic source predominates C_4–C_7: generated with depth (time).	Carbon isotopes, $\delta^{13}C$ reflect temperature of photosynthetic zone; -20 to -27 ‰
Concentrations of higher plants and animals much reduced; those of bacteria more important.	n-Paraffins; CPI near 1, range n-C_{20}–n-C_{27} Fatty acids; even CPI, predominate in n-C_{16}–n-C_1. range
Phytoplankton predominate; (with associated zooplankton).	

affins greater than n-C_{20} which exhibit no odd-carbon preference. Below some shallow but as yet indeterminate depth the full spectrum of light hydrocarbons is generated. It is only below this depth that full spectrum petroleum or incipient oils may exist.

The concept of a distinctive marine organic facies has several important implications for any theory on the origin of petroleum. In concluding, however, we stress only one of these—the similarity of most oils with the characteristics of the marine organic facies. Oils, extracts of rocks, and extracts from sediments of the marine organic facies generally lack an odd-carbon preference but do show carbon isotopic values that are similar or consistent for identical formations or source intervals. This suggests an environmental control. It may be that organic matter of marine origin is most important in the origin of most petroleum hydrocarbons; this could be true even when the source is not a true marine organic facies but merely a mixture of marine organic materials with much terrestrial organic detritus.

Literature Cited

(1) Smith, P. V., Jr., *Science* (1952) **116**, 437.
(2) Dunton, M. L., Hunt, J. M., *Bull. Am. Assoc. Petrol. Geologists* (1962) **46**, 2246.

(3) Rogers, M. A., Koons, C. B., *Trans. Gulf Coast Assoc. Geol. Soc.* (1969) **19,** 529.
(4) McIver, R. D., *Geochim. Cosmochim. Acta.* (1962) **26,** 343.
(5) Hunt, J. M., Jamieson, G. W., *Bull. Am. Assoc. Petrol. Geologists* (1956) **40,** 124.
(6) Cummins, J. J., Robinson, W. E., *J. Chem. Eng. Data* (1964) **9,** 304.
(7) Bray, E. E., Evans, E. D., *Geochim. Cosmochim. Acta.* (1961) **22,** 2.
(8) Welte, D. H., *Bull. Am. Assoc. Petrol. Geologists* (1965) **49,** 2247.
(9) Skopintsev, B. A., *Tr. Geol. Inst., Akad, Nauk SSSR* (1950) **17,** 29.
(10) Datsko, V. G., *Giorokhim. Materiale* (1959) **28,** 91.
(11) Bordovskiy, O. K., *Marine Geology* (1965) **3,** 5.
(12) Koons, C. B., Jamieson, G. W., Ciereszko, L. S., *Bull. Am. Assoc. Petrol. Geologists* (1965) **49,** 301.
(13) Davis, J. B., "Synthesis of Paraffinic Hydrocarbons by *Desulfovibrio desulfuricans,*" (abs.), *Geol. Soc. America* (1967) *Ann. Meetg., New Orleans,* Nov. 20–22, 1967.

RECEIVED May 14, 1970.

4

The Athabasca Oil Sands Development—50 Years in Preparation

C. W. BOWMAN

Imperial Oil Enterprises, Ltd., P. O. Box 3022, Sarnia, Ontario, Canada

Fifty years have elapsed since the first major surge occurred in the development of the Athabasca oil sands. The main effort has been devoted to the development of the hot water extraction process; 13 significant projects utilizing this process are reviewed in this paper. However, many other techniques have also been extensively tested. These are classified into several basic concepts, and the mechanism underlying each is briefly described. A critical review of K. A. Clark's theories concerning the flotation of bitumen is presented, and his theories are updated to accommodate the different mechanisms of the primary and secondary oil recovery processes. The relative merits of the mining and in situ *approaches are discussed, and an estimate is made of the probable extent of the oil sand development toward the end of this century.*

The past 50 years of the Athabasca oil sands' development are characterized by changing moods of great enthusiasm and deep frustration and discouragement. Many men have committed great portions of their lives to its development and, along with considerable success and professional acclaim, have experienced their share of public apathy and business skepticism. Two pioneers stand out in the literature on the sands—S. C. Ells and K. A. Clark. Their careers overlapped, both chronologically and professionally. Ells was the pioneering geologist, who spent many years studying the characteristics and behavior of various regions in the deposit. Clark was the experimentalist, the prime developer of the hot water separation method. His intuitive approach led him to an understanding of the mechanism of the separation process long before corroborating data were collected.

In the last four years, the rate of publication of in-depth research into specialized aspects of the oil sands has increased significantly. This reflects the progression of research from the "entrepreneurial" to the "consolidation" stage. An excellent review of the literature has been prepared by Kamp in the "Encyclopedia of Chemical Technology" (*1*).

In the present paper the various processes which have been proposed for separating the oil from the sands have been classified according to the key physical and chemical factors involved. This represents an application of concepts described previously by the author on the interfacial properties of the oil sands (*2*). In addition, the theories of K. A. Clark on the mechanism of the hot water process are updated in the light of new data on the flotation of oil.

Size and Scope of Deposit

Any report of the Athabasca oil sands requires some discussion of the magnitude of the deposit. The importance of the oil sands can be shown in their relation to two major Canadian resources—oil and minerals. Figure 1, a graphical representation of data from the *Oil and Gas Journal* (*3*), compares the oil sands with the conventional world petroleum reserves. Canada has less than 2% of the known reserves excluding the oils sands; when the latter are included, her share of the reserves increases to 36%. This comparison is biased in that the many other oil

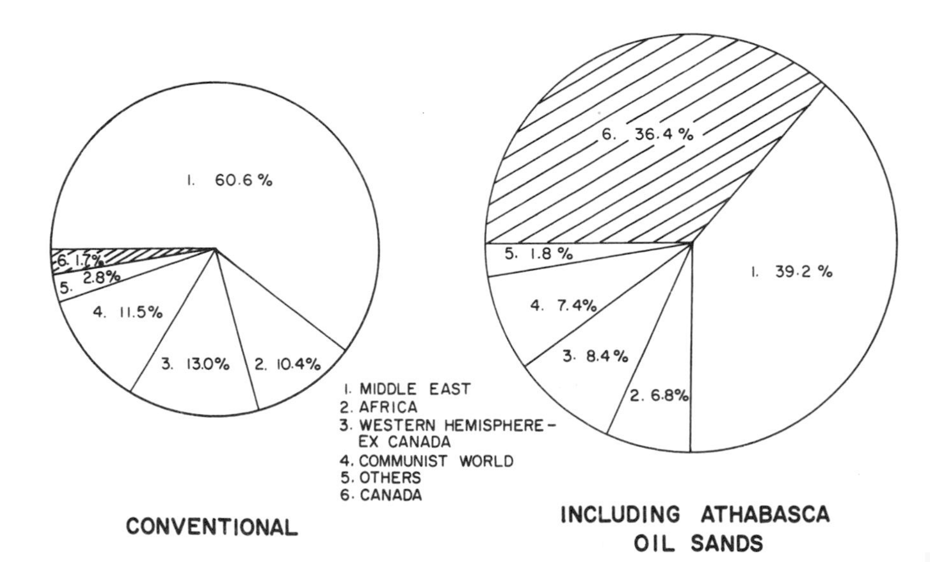

Figure 1. World petroleum reserves (based on data from the Oil and Gas Journal, *December 1969)*

sand and shale oil deposits in the world are excluded from the figures. However, it is still a useful comparison in relating the one oil sand deposit ready for development to what is now the existing competitive source of petroleum.

A mining or materials handling comparison is given in Figure 2. The amount of material mined in the existing commercial oil sand plant is approximately 50 million tons/yr. It has been estimated that by the time that the oil sand industry reaches a modest level of maturity, this will have increased to 500 million tons/yr. This is equivalent to building a St. Lawrence Seaway or Panama Canal every nine months.

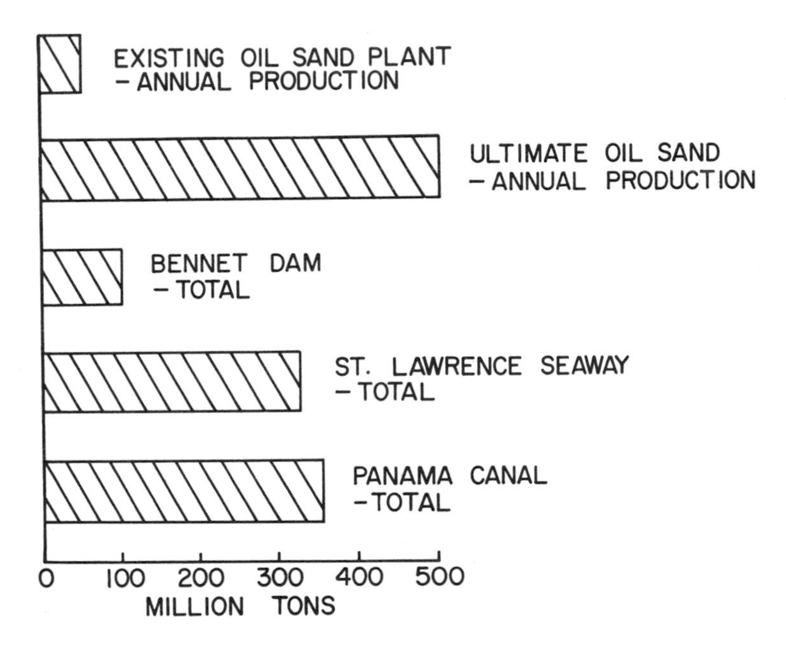

Figure 2. Material handling

It has been estimated that up to 20 wt % of the total deposit will be exploited by surface mining the sand and conveying it to a separations plant. The balance has too much overburden and will require *in situ* extraction. The discussion in this paper is confined largely to the mining route since it is the basis of the projects approved for commercialization to date.

Mining Approach

The hot water process has had the lion's share of the publicity during the past 30 years, but this is only one of several possible technologies for separating the bitumen from the sand. The various methods which have been proposed are grouped into five general types in Table I.

The first three methods involve water as the separating medium. These are based on the well known observation that the bulk of the sand, on the order of 99%, is not in direct contact with the bitumen but is "protected" by an envelope of water. On adding further water, the sand is "liberated" into the bulk water phase, and the bitumen, which was previously interspersed among the grains of sand, retracts into discrete flecks. These flecks are usually small, and having approximately the same density as water, remain dispersed in the water as an oil-in-water emulsion.

The first three methods in Table I differ in the manner in which the oil phase is separated from the water. In the hot water process, the bitumen flecks are attached to air bubbles to effect flotation.

In the second approach, a thick slurry of tar sand and water is kneaded to agglomerate the oil into balls, and these are separated by mechanical means such as screening. Four similar techniques for accomplishing this are recorded in the literature, the first in England in 1921. It is interesting that two of these methods were developed outside Canada.

The third approach is to add an organic solvent to decrease the density of the bitumen and thus allow a gravity separation from water and sand. In one case a centrifuge is used to enhance the rate of separation.

The final two techniques do not involve water as the separating media, and depend less on the surface chemistry considerations of the aqueous processes. The bitumen content of an oil sand is normally determined by extraction with an organic solvent, and a scale-up of this technique for commercial operation has been proposed.

Retorting of the total oil sand to produce cracked products was tested extensively, and other than the high heat loss to the sand, it appeared to be a very workable scheme. A combination scheme involving hot water washing to produce a wet bituminous froth followed by fluid coking of this froth was the basic operation recommended in the Blair report in 1950 (*4*).

In Situ *Approach*

Table I also includes a summary of four methods for recovering the hydrocarbon values without mining the oil sand. Three involve combustion as (at least) part of the displacement technology. The fourth uses a steam drive to emulsify the oil in water.

Hot Water Method

There have been at least 13 distinguishable efforts to develop a workable hot water separation process. These ranged in capacity (Table II) from small batch units to fairly substantial pilot plants, and culminated in the operation of a 45,000 bbl/day commercial plant.

The batch units operated by the Research Council of Alberta over a period of years yielded good quality primary bitumen products, containing as little as 5% mineral matter. The largest of the Research Council's pilot plants was constructed at Clearwater in 1929. It was a continuous unit which in general did not live up to the high froth quality standards of the smaller equipment. Clark attributed this to incomplete disintegration of the tar sand in the mixing step and excessive aeration of the tar sand slurry in its transport to the separation cell. He warned that the latter would be troublesome in the operation of large plants.

This difficulty in translating what appears to be a simple technique in a beaker to a viable continuous process contributed to the failure of many subsequent pilot plants. It was not until 1967 that the first commercial plant was put on stream, using giant bucket wheel excavators to mine the sand, the hot water process to separate the oil, centrifuging of the froth to remove solid and water contaminants, delayed coking of the bitumen to produce a sour distillate product, followed by hydrofining to produce a "synthetic" crude oil.

Mechanism of the Hot Water Extraction Processes

Clark prepared a series of statements in 1944 which summarized his view of the mechanism of the hot water process (*5*). In 1949 he revised these statements by proposing a new mechanism for flotation of the oil (*6*). In the 20 years which have elapsed, a number of papers have been published which include data and observations bearing on the validity of these statements. This would therefore appear to be a suitable time for a critical review of Clark's proposals and a formulation of revised statements where required. The two sets of statements are shown below. Statements 1 and 6 were largely unchanged in the two versions. Comments on the statements are given below.

K. A. Clark's "Statements" on Hot Water Process

Statement 1. Bituminous sand is an aggregate of sand, clayey matter, oil and water. The sand consists mainly of quartz particles of 50 to 200-mesh size and smaller, but also of particles of other minerals including mica, rutile, ilmenite, tourmaline, zircon, spinel, garnet, pyrite, and lignite. Clay occurs interbedded with the bituminous sand itself. Ironstone nodules of all sizes up to eight inches in diameter occur in the bituminous sand beds, especially in the southern part of the deposit. The oil is viscous, naphthenic, and of a specific gravity slightly greater than that of water. The oil content ranges up to, and sometimes exceeds 17% by weight. Rich bituminous sand from beds not invaded by water have a water content of 3–5% by weight. Some of the water is present as films on sand grains separating the oil from direct contact with the sand surfaces. Rich-looking bituminous sand is practically saturated with oil and water, differences in oil content being due mainly to differences in porosity of the mineral aggregate. The viscosity of the oil in the southern part of the deposit is many times greater than that of the oil in the northern part causing marked differences in the firmness of oil cementation of the bituminous sand beds in the two areas. The formation temperature of the bituminous sand deposit is about 36°F.

Statement 6. Satisfactory separation of oil from sand by the hot water process is impossible unless the natural packing of the bituminous sand is completely broken down in the pulping operation and unless the pulp is subsequently dispersed in excess water. The mineral particles and oil flecks must be free to move independently of each other under the influence of the small forces upon which the process depends.

Statements 2–5

1944 Version	*1949 Version*
(2). Water wets quartz and other siliceous minerals more readily than does mineral oil. Consequently water tends to displace the oil films surrounding the quartz and other siliceous particles when the bituminous sand is mixed with water. Whether the oil is completely displaced depends on the properties of the water. That is to say, the materials dissolved or suspended in the water modify its wetting properties.	(2). Water wets quartz and other siliceous minerals more readily than does mineral oil. Consequently, when bituminous sand is mixed with water, the already existing separation of oil from sand surfaces by water is at least preserved, and may be extended if not already complete.
(3). When bituminous sand is mixed with water, substances present dissolve or become suspended in the water and determine its ability to displace the oil from the sand surfaces. This ability is a function of the concentration of substances dissolved in or suspended in the water.	(3). When bituminous sand is mixed and heated with water into a mortar-like pulp, the oil is dispersed into small oil flecks which lie unattached among the sand grains. The content of clayey material in bituminous sand plays an important role in fleck formation and fleck formation plays an important role in the hot water separation process. In the rare cases of bituminous sands containing practically no clay, the separation process proceeds unsatisfactorily.
(4). When bituminous sand is mixed with a small quantity of water, the concentration of substances which dissolve or are suspended in the water is greater than when a large quantity of water is used. The small quantity of water, after having been mixed with the bituminous sand, is a better wetting agent and displaces the oil from the sand more completely than is the case with a large quantity of water.	(4). The oil flecks present in a bituminous sand pulp vary in size and in the amount of clayey material that is associated with them. As the clay content of bituminous sand increases, the amount of clay associated with the oil flecks increases.

Statements 2–5 (Continued)

1944 Version	*1949 Version*
(5). Bubbles of air or of water vapor present in a system of water, mineral matter and small masses of oil, will become attached to the oil masses and will float them to the surface. However, only small bubbles a few millimeters and less in diameter are effective in floating the oil. Air and water vapor bubbles also become attached to and float mineral particles. Particles of minerals other than quartz are floated more readily than quartz.	(5). When a hot bituminous sand pulp is flooded with hot water under conditions of agitation and access to air, oil flecks of low clay content gather together into a froth. This froth occludes or otherwise retains sand particles in amount depending on conditions. Particles of minerals other than quartz are retained more readily than are quartz particles. Oil flecks of high clay content are not gathered into the froth and remain dispersed in the sand and water. The clay content which determines whether an oil fleck will go with the froth or not varies with conditions but is in the neighborhood of 10%. Oil flecks which become dispersed in the water can be induced to form froth by submitting them persistently to conditions which bring the flecks in contact with the water–air interface.

Statement 1. This is still an accurate description of the oil sand system. The clay is largely kaolinite and illite, with essentially no evidence of montmorillonite (2). The viscosity of the bitumen ranges from 5000 to about 75,000 cS at 100°F, with the higher values being observed near the bottom of the deposit (2). As the bitumen content of an oil sand increases, its water and clay content decrease (7).

Statements 2 to 5. These were modified appreciably in the 1949 revision. In the earlier version, Clark believed that the properties of the water phase were very important in oil–sand separation, and he emphasized the role of air in floating the bitumen particles. In the later version, both these variables were given less emphasis. The critical factor was believed to be the association of clay with oil to form the somewhat mysterious "flecks," which formed froth by submission "to conditions to bring the flecks in contact with the water–air interface." In the body of the 1949 paper Clark left no doubt that convection currents were the mechanism not flotation in the normal sense of air–oil attachment. Clark backed up his new theory with some very revealing tests on simple clay–oil dispersions. Before discussing the merits of the two sets of statements, it would be of value to examine new data not available to Clark at that time.

Table I. Comparison of

Process	*Basis of Separation*
Mining Approaches	
Hot Water	Displacement of oil by water; attachment of oil to air bubbles to effect flotation.
Oil Agglomeration	Kneading of tar sand with water to agglomerate oil into balls; separation of oil balls by screening, etc.
	(a) Fyleman process
	(b) Spherical agglomeration (NRC[a])
	(c) Phase exchange
	(d) Sand reduction (Esso)
Combined Solvent/ Water Processes	Contact of tar sand with hydrocarbon diluent; dispersion of mixture into water to effect density separation.
	(a) Cold water process (Canadian Dept. of Mines)
	(b) Fluidizing Water Wash (Cities Service)
	(c) Centrifugation (Can-Amera)
Solvent Extraction	Extraction of the bitumen by a hydrocarbon diluent; separation of oil phase by filtration. (Cities Service)
Fluid Coking	Retorting of the tar sand in a fluidized bed
In-Situ Approaches	
Forward Combustion	Injection of hot air into deposit to effect distillation and thermal cracking (Pan American Petroleum)
Combined Combustion/ Steam Drive	Injection of hot air into production well to promote combustion; steam injected into producing well to reduce viscosity prior to putting the well on production. (Pan American Petroleum)
Steam Drive	Injection of hot water and steam into deposit to displace oil and emulsify it in water; caustic used to increase efficiency (Shell Oil)
Nuclear Explosion	Thermal cracking of oil by nuclear energy followed by condensation and drainage into cavity.

[a] Abbreviations: NRC = National Research Council; C.P. = Canadian Patent;

Bitumen Separation Processes

Principal Developers	*Principle References*
K.A. Clark D. S. Pasternak	*Can. Oil Gas Ind.* (1950) **3,** 46. *Proc. Athabasca Oil Sands Conf.* (1951) 200.
E. Fyleman	*J. Soc. Chem. Ind.* (1922) **41** (2), 14T. C. P.[a] **203,676** (1922).
I. E. Puddington	*Petro Process Eng.* (Oct. 31–35, 1963). C. P. **787,898** (1963).
E. Weingaertner	*Erdol Kohle* (1960) **13,** 549.
J. A. Bichard	K. A. Clark Volume, pp 171–191, Research Council of Alberta, 1963. C. P. **675,912** (1963).
L. E. Djinghenzian	*Proc. Athabasca Oil Sands Conf.* (1951) 185.
J. D. Frame J. D. Haney E. W. White	C. P. **639,713**
G. R. Coulson	C. P. **491,955**; **593,381**; **596,561**; **602,087**
J. H. Cottrell	"K. A. Clark Volume," pp. 193–206 Research Council of Alberta, 1963. U.S.P.[a] **3,117,922** (1964).
P. E. Gishler W. S. Peterson	*Can. J. Technol.* (1956) **34,** 104. 1956. C. P. **530,920**
L. E. Elkins	C. P. **546,390**
K. L. Hujsak H. Grekel R. Mungen	U.S.P. **3,384,172**
J. Offeringa T. E. Doscher	U.S.P. **3,385,359** "K. A. Clark Volume," pp. 123–141, Research Council of Alberta, 1963.
M. L. Natland	"K. A. Clark Volume," pp. 143–155, Research Council of Alberta, 1963.

U.S.P. = U.S. Patent.

(*a*). The substances which dissolve in water generally tend to produce an alkaline system, as required for good separation. Oil sands near the surface of the deposit occasionally produce an acidic slurry and are susceptible to a further pH decrease when exposed to the atmosphere. Values as low as 3 have been observed. This lowering of pH is probably caused by oxidation of siderite and sulfur containing minerals (*2*).

Table II. Hot Water Extraction Development

Operation	*Period of Operation*	*Maximum Capacity, bbl/day*
Research Council of Alberta at University of Alberta, Edmonton	1923	Batch
Research Council of Alberta at Dunvegan Yards, Edmonton	1924–1925	15
Research Council of Alberta, Edmonton	1925	Batch
Federal Dept of Mines and Research Coun- of Alberta at Clearwater, Alberta	1929–1930	40
International Bitumen Company (R. G. Fitzsimmons) at Bitumount, Alberta	1930–1942	200–350
Abasand Oils Limited (M. W. Ball → Federal Government) Denver → Toronto → Abasand, Alta	1936–1945	350
Consolidated Mining and Smelting Co. of Canada, Ltd., Chapman, British Columbia	1939	Batch
Alberta Government Plant, Bitumount, Alberta	1949–1950	500
Can-Amera, Royalite, Bitumount, Alberta	1955–1957	(Same basic plant as one above)
Cities Service Athabasca, Inc., Mildred Lake, Alberta	1959–1963	750
Syncrude Canada Ltd., Edmonton, Alberta	1963–present	15
Great Canadian Oil Sands, Tar Island, Alberta	1965	150
Great Canadian Oil Sands, Tar Island, Alberta	1967–present	45,000

(*b*). The substances also lower the surface tension of the water to 60–65 dynes/cm. A low surface tension retards the attachment of air bubbles to both solid and oil particles, which are competing processes. The flotation of oil alone is desired. However, without knowing the changes in the other interfacial tensions (solid–water, solid–air, and oil–water, oil–air), one cannot state categorically the effect of a low aqueous surface tension. This is discussed more fully elsewhere (*2, 8*).

(*c*). To achieve the maximum lowering of surface tension, no more than one part of water should be added to one part of oil sand (*2*). However, the effect of water addition on pulp viscosity is very pronounced, and any beneficial effects of low water are probably caused by more effective disintegration of the oil sand, and more effective air–bitumen contact, with a thick slurry (higher shear rate).

(*d*). High speed photographs have been taken of the flotation process (Figure 3) which show clearly conveyance of bitumen particles to the water surface by attached bubbles. Analysis of the bubbles indicates they are largely air (and water vapor), not hydrocarbon gases. It is believed that this air attachment occurs during the pulping step, when there is sufficient mechanical energy to overcome the similar electrical

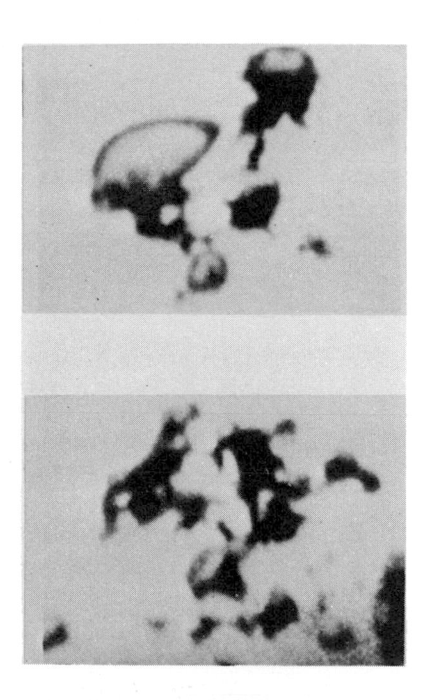

Figure 3. High speed photographs of bitumen flecks conveyed to surface by air bubbles

charges on both the bitumen particles and air bubbles. The attachment of air bubbles proceeds by the following mechanism:

(1) Adsorption of surfactants (predominantly naphthenic acids) at the bubble surface
(2) Contact attachment of bubble to bitumen
(3) Spreading of bitumen around the bubble to complete attachment

In assessing this information, the author believes that Clark's original statements were closer to describing the mechanism for flotation of bitumen in the primary separation step. There appears to be no doubt that conveyance of bitumen to the surface by attached bubbles is an important factor. For the 10–20% of the bitumen which is not recovered in this operation, the role of convection currents is more dominant. This "secondary" oil is truly in the form of "flecks," and is highly dispersed in the aqueous middlings. These can be brought to the air–water surface by agitation, and high speed photographic observation substantiates Clark's belief that this is not by the attachment of bubbles (at least visible bubbles).

The above is not to underestimate the role of clay in the hot water process. Tar sands of high clay content generally are difficult to process, owing partly to the tendency of clay to stabilize oil and water emulsions. In addition, polyvalent metal ions act as bridging agents between the oil and clay, particularly at low pH. The association of clay with the oil retards the attachment of air as well as increasing the density of the oil particles. This mechanism is discussed in detail in Ref. 2.

Statement 6. This is a fairly obvious comment, but it has serious implication in the selection of the slurrying time. Both the lump size and temperature vary in the feed to the slurrying unit, and the selection of an optimum time is difficult. Useful experimental data were presented by Great Canadian Oil Sands in 1967 (7).

The Athabasca Oil Sands—An Exploitable Resource

The main effort in oil sand development has been directed to the separation of the bitumen for subsequent refining. However, this was not always the goal. Several attempts were made between 1910 and 1930 to use the total oil sand as a paving material, with a fair measure of success. Stretches of experimental pavement lasted for many years.

The future development of the oil sands will take into account other products or services which are directly or indirectly associated with the deposit. The sand itself is composed of many minerals which have industrial significance, as indicated in Table VI of Ref. 2. An integrated network of primary and secondary industries (based on both inorganic and organic products) will one day arise to make the Athabasca valley one of the industrial centers of Canada.

Literature Cited

(1) Kamp, F. W., "Encyclopedia of Chemical Technology," 2nd ed., Vol. 19, pp. 682–732, Interscience, New York, 1963.
(2) Bowman, C. W., "Molecular and Interfacial Properties of Athabasca Tar Sands," *Proc. World Petrol. Congr., 7th,* 1967.
(3) *Oil Gas J.* (1969) **67** (52) 94.
(4) Blair, S. M., "Report on the Alberta Bituminous Sands," Government of the Province of Alberta, Edmonton, 1950.
(5) Clark, K. A., *Trans. Can. Inst. Mining Metall.* (1944) **47**, 257.
(6) Clark, K. A., Pasternak, D. S., "The Role of Very Fine Mineral Matter in the Hot Water Separation Process as Applied to Athabasca Bituminous Sand," *Res. Council Alberta, Rept.* **53** (1949).
(7) Innes, E. D., Fear, J. V. D., *Proc. World Petrol. Congr., 7th,* 1967.
(8) Leja, J., Bowman, C. W., *Can. J. Chem. Eng.* (1968) **46**, 479.

RECEIVED June 8, 1970.

5

Catalytic Reforming

EUGENE F. SCHWARZENBEK

Process Consultant, 132 Main St., Chatham, N. J. 07928

The role of the catalytic reforming process in meeting the future demand of high octane lead free gasoline is discussed. An increase in the severity of reforming is needed to produce aromatic-rich fractions with octane numbers as high as 110. Operating variable effects illustrate the need to reform at low pressure to minimize the loss of valuable liquid to by-product gas and butane. The new platinum–rhenium catalyst shows improved stability and will allow the design of low pressure semiregenerative systems without the requirement of high investment regeneration facilities. There still is, however, considerable potential in the improvement of the reforming catalyst by increasing its dehydrocyclization activity and decreasing its hydrocracking activity.

The catalytic reforming process for converting low octane naphtha to high octane gasoline and aromatic hydrocarbons for petrochemical use continues to have a history of dramatic improvements in catalyst formulation and processing technique. Within the past year there has been both a new catalyst and a new process announced. Chevron introduced a new platinum–rhenium catalyst which exhibits significantly better aging characteristics than the conventional platinum catalysts (*1, 2*). Englehard introduced the Magnaforming process which has a modification in the process flow of the recycle gas and gives higher liquid yields than obtained with the conventional recycle gas flow (*3*).

All catalytic reforming processes are receiving wide attention as a result of recent public and governmental attention to the problem of air pollution from the automobile engine. The impending legislative limits on the lead content of gasoline will make catalytic reforming the principal refining process for producing high octane gasoline.

History

The development of catalytic reforming has been primarily a problem of controlling coke deposition and catalyst aging. The original molybdena–alumina reforming catalyst aged rapidly, and short cycle fixed bed or continuous fluid bed regeneration was practiced. However, the introduction of a platinum catalyst by Universal Oil Products in a nonregenerative type operation had a significant effect on the economics of catalytic reforming, and soon reforming development work by the many oil research organizations was centered on the use of this type catalyst. The result was many processes using various platinum catalyst formulations and processing conditions. A list of the installed U.S. and Canadian capacities as of January 1, 1970 is given in Table I. Capacity in the United States is equivalent to about 22% of crude capacity, and in recent years it has been expanding at the rate of 8% per year compared with 5% for crude capacity expansion.

Despite the many competitive processes being offered to the petroleum refining industry, the Platforming process still has a major portion of the presently installed capacity. This process was introduced as nonregenerative, operating at a high recycle gas rate and at a pressure of at least 750 psig. The trend in operating pressure of all catalytic reformers has been downward, though the bulk of today's operating capacity is in the 450–500 psig range. Processes other than Platforming have stressed the regenerability of the catalyst. The platinum catalysts are difficult to regenerate, but good development work has resulted in techniques and conditions which restore a substantial portion of the reforming characteristics of the catalyst. Advantage is taken of the ability to regenerate

Table I. Catalytic Reforming Capacity BPSD Jan. 1, 1970

Country	*United States*	*Canada*
Crude Capacity	*12,651,375*	*1,438,749*
Platforming	1,286,760	132,000
Ultraforming	297,190	
Englehard	249,740	6,400
Powerforming	211,800	75,000
Mobil	194,800	
Houdriforming	148,450	5,600
Catforming	81,700	
Miscellaneous	305,780	15,000
Total	2,776,220	234,000
% on Crude	22	16
Growth rate 1967–1970		
Crude capacity	4.9	6.0
Reformer capacity	8.3	6.3

the catalyst by operating at lower pressures and thus increasing liquid product yields. The trend in the design of semiregenerative processes has been to operate in the range of 350 psig. The Ultraforming and Powerforming processes are cyclic type regenerative processes operating for relatively short on-stream periods. The catalyst is rugged, and a regeneration-rejuvenation technique has been developed which effectively restores the catalyst's reforming characteristics. These processes, as a result, take full advantage of the high yields at low pressures and operate at about 200 psig with low recycle gas rates. The cyclic type units are high in capital investment, but they can reform naphthas effectively to high octane levels.

In May 1969 Chevron announced a major new catalytic reforming development in the form of the platinum–rhenium catalyst. The introduction of this new catalyst is particularly opportune in view of Federal pollution legislation on the emission of particulate matter from automobile exhausts and the eventual elimination of lead from all gasoline. These regulations will place emphasis on the production of high octane lead free gasoline by the catalytic reforming process. The new catalyst will be of substantial benefit to the oil refining industry in that it will allow the operation of existing reformers and the design of new semiregenerative reformers at much severer operating conditions than were possible in the past.

Automotive Pollution

Typical Federal emission standards presented in Table II cover evaporation control and the emission of hydrocarbon, carbon monoxide, oxides of nitrogen, and particulate matter from the automobile engine exhaust. These specifications are based on a standard car testing procedure involving acceleration, deceleration, idling, etc. The particulate matter has been converted from a gram/mile basis to an equivalent amount of lead as grams/gallon of gasoline using as a conversion basis the average U.S. consumption of gasoline/mile.

Table II. Federal Emission Standards for a 4000 lb Automobile

	Prior Control	*1970*	*1975*	*1980*
Hydrocarbon, grams/mile	11.4	2.2	0.6	0.25
Carbon monoxide, grams/mile	82.6	23	11.5	4.7
Nitrogen oxides, grams/mile	3.9	—	0.95	0.4
Evaporation control	none	—	yes	yes
Particulate, grams/mile	0.25	—	0.10	0.03
Lead, grams/gal equiv.	2.4	—	1.0	0.3

Emission studies show that lead is only a small part of the automotive pollution problem. Prior to control, hydrocarbon emissions were more than 40 times and the oxides of nitrogen emissions more than 15 times the emission of the lead compounds. Obviously, however, legislation will result in the eventual elimination of lead from gasoline. The removal of lead, besides eliminating a possible toxic pollutant, simplifies the problem of handling other automotive exhaust pollutants in that catalytic exhaust chambers perform much better in the absence of lead contaminant. All emission standards become particularly severe in 1975 and 1980. The particulate standards are equivalent to 1 gram Pb/gal in 1975 and 0.3 gram Pb/gal in 1980. Since the particulates include all solid materials, tolerable lead levels will be less than indicated above.

Present Day Gasoline Pool

The process and investment requirements and increased operating costs for producing lead free gasoline equivalent to present day gasoline in octane quality was the subject of an extensive API sponsored report prepared by Bonner & Moore Associates, Inc. (*4*). The analysis was based on the petroleum refining industry operation in 1965. Several refinery models were used to simulate the industry according to size and location. The report showed that high investments and drastic changes in processing, particularly in the area of catalytic reforming, were required to produce high quality lead free gasoline. An indication of the problem can be obtained from an analysis of the composition and octane qualities of the present day gasolines and their various components (Table III).

Present day premium gasoline contains primarily catalytic reformate, cat cracked gasoline, and alkylate. Regular gasoline contains, in addition to cat reformate and cat cracked gasoline, light straight run, natural gasoline, and the thermal gasolines from coker, visbreaking, and thermal cracking operations. These latter gasolines are low in octane quality but with the addition of lead can be incorporated in regular gasoline.

Inspection of the octane qualities of the various components of the present day gasoline pool show none of the components to have a research octane number on a lead free basis higher than that for premium gasoline. Cat cracked gasoline and alkylate, which are major components of the gasoline pool, cannot be changed significantly in octane quality by changing process conditions. The quality of cat reformate, on the other hand, is subject to control by changes in operating conditions. Whereas the average present day operation is at about 91 octane number research clear, octane numbers above those for present day premium gasoline can be obtained. Since aromatics are the only compounds with research octane numbers above 100, the production of high quality lead free gasoline

Table III. U.S. Gasoline

	Composition, vol %		
	Premium	*Regular*	*Pool*
Butanes	5.9	4.0	4.7
Naphtha	1.4	27.1	17.6
Reformate A	4.0	25.1	17.2
Reformate B	12.2	0.5	4.8
Cat cracked	51.5	36.3	41.9
Alkylate	25.0	—	9.1
Thermal gasoline	—	4.2	3.0
Raffinate	—	1.6	1.0
Polymer	—	1.2	0.7
	100.0	100.0	100.0
TEL, cc/gal	2.7	2.3	2.4

depends primarily on the production of these materials by catalytic reforming.

Lead Free Gasoline Composition

Compositions of lead free gasoline with octane qualities equivalent to present day gasoline are shown in Table IV. These compositions represent the sum total of the many different refinery models presented in the API report.

The low octane straight run and thermal gasolines have been essentially eliminated from the gasoline pool. Alkylation has increased only slightly, though there has been a significant shift to amylene alkylation. Pentanes from cracking and from crude have been isomerized to isopentane. There has been a significant increase in cat reformate from 22% for the leaded gasoline pool to 43% for the lead free pool. The average octane number of the cat reformate is about 103 research clear, though there is considerable splitting and extraction to produce aromatic concentrates with research octane numbers over 110 for use in the premium grade of gasoline. The amount of cat cracked gasoline has decreased from 42% on a leaded basis to only 22% on a lead free basis. This is not a result of decreased cat cracking gasoline production but is attributable to secondary treatment of the gasoline—alkylation of amylenes, isomerization of *n*-pentane, and cat reforming of the heavy cat gasoline.

The olefin content of the gasoline pool has decreased from 19% on a leaded basis to only 9% on a lead free basis. Aromatics, however, have increased from 21 to 42%, respectively, and are as high as 51% in the premium gasoline.

Pool—1965 Refining Basis

Research		Motor	
clear	*3cc TEL*	*clear*	*3cc TEL*
96	104	98	104
73	90	73	91
90	98	81	90
95	101	85	92
93	98	80	85
93	105	92	105
74	87	66	77
62	83	62	83
93	99	81	85
.... 100		91	
.... 94		86	
.... 96.5		88.5	

Table IV. Lead Free Gasoline—1965 Refining Basis

	Composition, vol %			*Octane clear*	
	Prem.	*Reg.*	*Pool*	*Rsh.*	*Motor*
Butanes	5.7	3.4	4.3	96	98
Isopentane	15.9	13.6	14.4	93	88
Reformate A	—	6.4	4.0	100	89
B	2.2	20.8	13.9	103	92
C	16.5	—	6.2	105	94
Light reformate	—	9.1	5.8	91	84
Heavy reformate	16.2	—	6.1	110	100
Aromatic ext.	16.6	—	6.2	110	100
Cat cracked	10.7	29.4	22.3	94	81
Alkylate	13.8	11.1	12.0	93	92
Hydrocrackate	0.8	3.5	2.5	85	82
Miscellaneous	1.6	2.7	2.3		
	100.0			101	90
		100.0		95	85
			100.0	97.5	87
Olefins	7	11	9		
Aromatics	51	36	42		

Cat Reformate Splitting

The liquid product from catalytic reforming is essentially a blend of aromatics and paraffins, the aromatics concentrating in the high boiling portion and the paraffins in the low boiling portion of the gasoline. The high octane aromatic concentrates required for premium gasoline blending can, as a result, be produced by distillation as well as by solvent extraction. Typical qualities of light and heavy reformate fractions are presented in Table V. Heavy reformates with aromatic contents greater than 95% and research octane numbers of 110–111 can be obtained.

Table V. Cat Reformate Splitting

C_5 + reformate					
Rsh. oct. clear	95	100	103	105	95
Aromatics, vol %	52	63	80	90	52
Separation			Splitter		Extract
Heavy reformate yield, vol %	60	57	55	53	56
Rsh. oct. clear	103.4	108.0	110.2	111.2	110.2
3cc TEL	106.8	111.4	113.6	114.6	113.6
Mot. oct. clear	92.9	98.0	100.2	101.2	100.2
3cc TEL	96.9	102.0	104.2	105.2	104.2
Light reformate					
Rsh. oct. clear	77.0	83.1	87.7	91.6	61.6
3cc TEL	91.3	92.5	94.7	96.7	83.4
Mot. oct. clear	72.0	77.1	81.3	84.8	62.6
3cc TEL	82.2	82.5	84.0	86.3	83.6
Aromatics, vol %					
Heavy reformate	—	—	95	98	95
Light reformate	—	—	62	80	4

Lead Free Gasoline Costs

The increased investment requirement and increase in gasoline cost for producing lead free gasoline with today's octane quality is presented in Table VI. The total process investment requirement was 4.2 billion dollars with 50% of the total being associated with catalytic reforming—reaction, reformate splitting and extraction, and feed pretreatment. Catalytic reforming capacity will have to be doubled to meet the requirements of lead free gasoline with today's quality. The average increase in gasoline processing costs was 2.2¢/gal. Since high octane lead free gasoline will depend strongly on the production of aromatics, the value of these materials for petrochemical use will also increase. This was calculated to be 12¢/gal.

The distribution of investment monies spent by the oil industry in 1965 is shown in Table VII. The above process investment requirement for producing high quality lead free gasoline is about seven times the

Table VI. Investment Requirements and Lead Free Gasoline Cost for the U.S. Refining Industry

Process Investment	*MM/*	*%*
Aromatics	2,108	49.8
Cracking	1,098	25.9
Isomerization	376	8.9
Alkylation	293	6.9
Miscellaneous	358	8.5
Total	4,233	100.0

Gasoline Refining Cost[a]	*¢/gal*
Small Refiner	4.7
Large Refiner	1.8
Average	2.2

[a] Incremental cost over 1965 leaded gasoline basis.

amount normally spent by the petroleum industry for oil refining plants. Furthermore, the monies spent by the petroleum industry on oil refineries are only a small fraction of the total spent to bring their products to the market place. It is therefore expected that the costs of lead free gasoline to the consumer will be considerably higher than the 2.2¢/gal shown above for refining costs alone.

The eventual octane levels reached for lead free gasoline cannot be predicted easily at this time since the production of these high quality products could result in severe economic pressures on the oil refining industry. The initial and ultimate octane quality will be a compromise between the higher cost of producing lead free gasoline and the loss in performance of the automobile engine. The history of engine compression ratio and gasoline octane quality is shown in Table VIII. Complete removal of lead from present day gasoline without significant processing changes would result in gasoline quality equivalent to that in 1952 when new car compression ratios were on the average 7.2. Since the automobile industry is contemplating 8.5 compression ratio cars to operate on lead

Table VII. Petroleum Industry Investment in 1965

	MM$
Production	4,370
Pipe lines	265
Refineries	600
Chemical plants	525
Marketing	1,000
Other	225
	6,985

Table VIII. Automobile Compression Ratio and Gasoline Octane Quality

Year	*Research Octane*			*New Car*
	Prem.	*Reg.*	*Pool*	*CR*
1965	100	94	96	9.4
1960	99	92	94	8.9
1956	96	89	91	8.5
1955	95	87	89	8.0
1952	91	84	86	7.2
1950	89	83	85	7.0
1965 Lead free	91	84	86	

free gasoline, the octane quality of today's gasoline pool before addition of lead will have to be raised to meet this demand. Current and future catalytic reformers will be required to operate at much severer operating conditions than presently being used. An 8.5 compression ratio was the average for new cars in 1956 when the octane quality of the total gasoline pool was 91.

Reformer Octane Potential

One of the major causes for the increased costs of lead free gasoline is the loss in liquid product yield when catalytic reforming to high octane levels. The effects presented in the API report are shown in Table IX. Although current reformers operate over the range 200–750 psig, most of the units operate at 450–500 psig and a weight space velocity of 2.8. These units were generally designed for low octane levels but are considered to have economic capability to produce 95 octane number product from an average type naphtha. However, when pushed to octane levels as high as 100, liquid yield losses are excessive.

Table IX. Catalytic Reforming Low and High Severity Yields of Naphtha (N + 2A = 60)

Rsh. clear	*85*	*90*	*95*	*100*	*103*	*105*
Current Reformers						
pressure	450–500					
w/hr/w	2.8					
vol % C_5 +	86.5	83.0	75.0	55.0	—	—
High Severity Reformer						
pressure				350		
w/hr/w				1.5	1.0	0.8
vol % C_5 +	86.5	83.8	80.0	74.5	71.1	67.6

The trend in reformer design has been to decrease operating pressure and space velocity when producing high octane gasoline. At the time of the API report, a high severity reformer operating at 350 psig and low space velocity was considered capable of producing 105 research octane reformate. Yield losses for the high severity reformers were considerably less than from the current high pressure reformers. To meet the average demand of 103 octane reformate when producing lead free gasoline, the current high pressure reformers were limited to the processing of high quality feed stocks, such as heavy hydrocrackate and hydrotreated heavy cat cracked gasoline, or they were combined with solvent extraction to produce high octane aromatic concentrates.

Catalytic Reformer Pressure

The optimum design pressure of a catalytic reformer is a balance of the lower investment and lower liquid yields at high pressure and the higher investment but higher liquid yields at low pressure. The balance depends in particular on the amount of investment monies associated with regeneration. Considerably less money is involved in a semiregenerative operation in which the unit is shut down and the catalyst regenerated after an extended period of operation than in a continuous system where there is frequent shutdown and shifting from one reactor bed to another. With the conventional platinum reforming catalysts, a 350-psig pressure usually gave sufficiently slow aging rates that long operating periods could be used and only infrequent regeneration was necessary. At 200-psig pressure, however, aging in general was sufficiently severe that frequent regeneration was necessary. The new platinum–rhenium catalysts, which have exhibited appreciably lower aging rates, make it possible to design a semiregenerative system at a lower pressure and thus take advantage of an improved product distribution—higher hydrogen production and higher liquid product yields. Careful analysis of the role of pressure on economics is therefore required to arrive at the optimum design.

Octane improvement in catalytic reforming is a result of a dehydrogenation reaction to form aromatics and a hydrocracking reaction to convert higher boiling paraffins to lower boiling higher octane material. Pressure is an operating variable which controls the relative amounts of these two reactions. Hydrocracking is promoted by high pressure, dehydrogenation by low pressure. A study of the effect of pressure on product distribution shows essentially the same effect on dry gas and butane production for a range of reforming severities (Figure 1). Increasing pressure from 350 to 500 psig increases gas and butane production 25 to 40% when operating at the same octane level. Decreasing the pressure from 350 to 200 psig results in decreases of 20 to 30%. The effect of pressure was more pronounced on the lower boiling feeds.

Typical effects of pressure on dry gas, butane, C_5+ liquid, and hydrogen production are shown in Figures 2 and 3 for the reforming of a

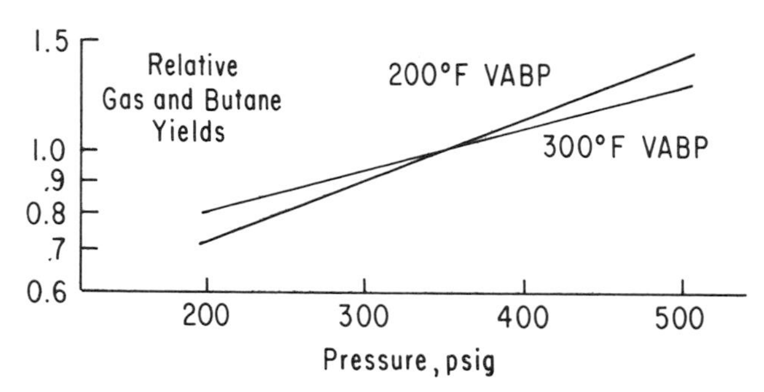

Figure 1. Effect of pressure on gas and butane production

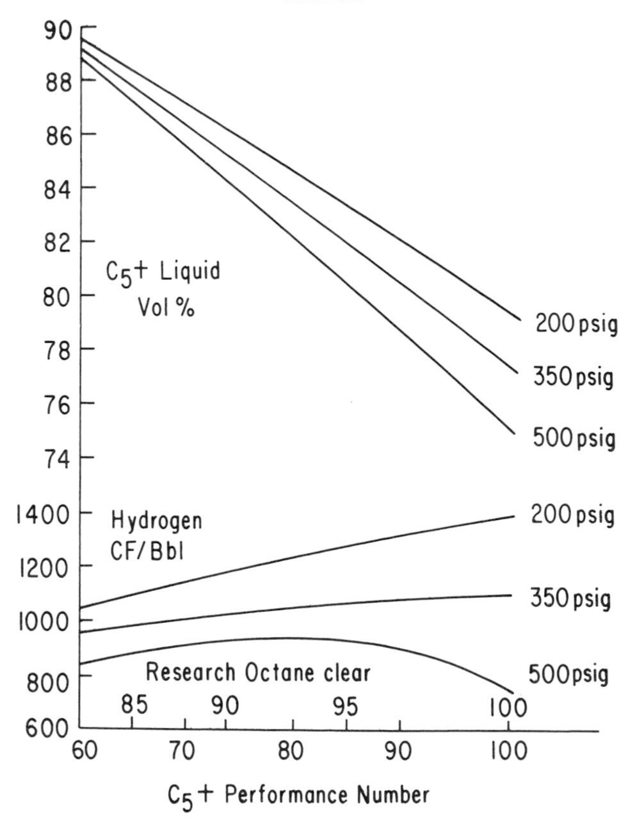

Figure 2. Catalytic reforming of a 54°API Mid-Continent naphtha

Mid-Continent naphtha over a range of octane qualities. The effects of pressure become more significant at the higher octane levels. Of particular interest is the effect of pressure on hydrogen production. As reforming severity is increased at 500 psig pressure, hydrogen production increases rapidly at low octane levels, reaches a peak production, and then decreases at high octane levels. The lack of any significant increase in hydrogen production at high octane levels is an indication of the extensive amount of hydrocracking occurring under these conditions. Low pressures decrease hydrocracking and increase hydrogen production. Hydrogen production and hydrogen content of the recycle gas are both good and sensitive indices of the selectivities of the operation. Generally speaking, an increase in liquid product yield at a given octane quality will be accompanied by an increase in hydrogen production to satisfy the carbon–hydrogen balance of the system.

The effects of pressure when reforming paraffinic and naphthenic naphthas at 100 research octane number are given in Table X. The lower pressure gives a higher incremental gasoline yield with the more

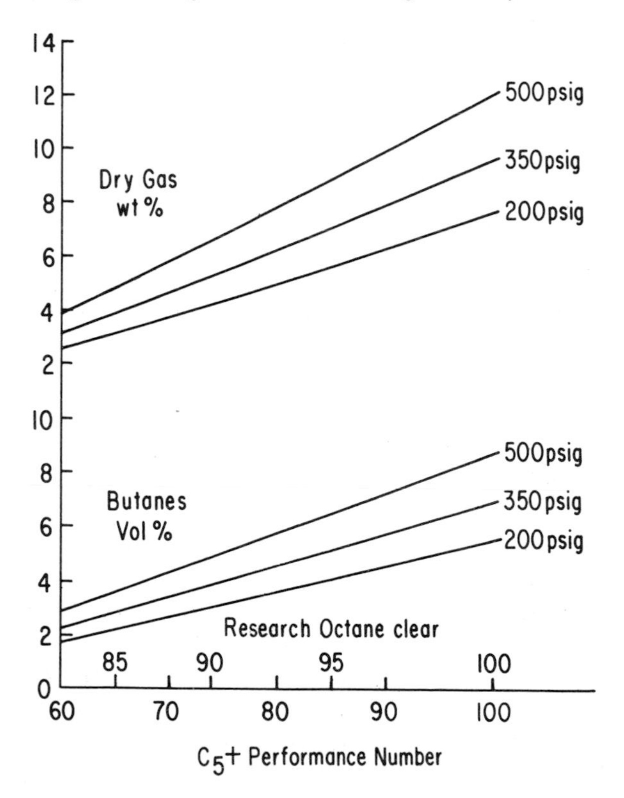

Figure 3. Catalytic reforming of a 54°API Mid-Continent naphtha

Table X. Effect of Pressure on Product Distribution

Feed	*Mid. Cont.*			*W. Texas*	
Gravity °API	54			52	
VABP, °F	300			285	
% Naphthenes	41			38	
% Aromatics	8			18	
% Paraffins	51			44	
Research clear	100				
Pressure, psig	200	350	500	200	350
Hydrogen, CF/bbl	1390	1100	750	1310	1080
Dry gas, wt %	7.6	9.6	12.1	6.0	7.6
Butanes, vol %	5.5	6.9	8.7	4.1	5.2
C_5 + liquid, vol %	79.5	77.5	75.2	83.0	81.5
C_5 + liquid RVP	3.4	3.8	4.1	3.2	3.6

paraffinic naphtha. In all comparisons, the lower pressure gives a lower vapor pressure product and would, as a result, show a greater advantage in liquid product yield when evaluated on a 10 lb RVP basis.

Catalyst Aging

The product distributions previously discussed were on the basis of fresh catalyst or beginning of run conditions. The selectivity characteristics of most platinum catalysts become poorer with use. The catalyst loses dehydrogenating activity as a result of coke deposition or physical changes in the composition or structure of the catalyst. The loss in dehydrogenation activity is greater than the loss in hydrocracking activity. As a result, when temperature changes are made to compensate for activity loss, there is proportionally more hydrocracking. Gas yields increase, and hydrogen production and C_5+ liquid yield decrease.

Temperature rise required to maintain a given octane quality of the product has been taken as an index of catalyst activity loss and has been related in Figure 4 to loss in catalyst selectivity measured by absolute liquid yield loss and by percentage increase in dry gas and butane yields. The relationships serve to illustrate one of the most significant differences getween the conventional platinum and the new platinum–rhenium catalysts. The new rhenium activated catalysts have outstanding stability with regard to product distribution selectivity, this stability being maintained at high and low pressures and at high octane levels. Whereas a 50°F rise in temperature during the process cycle would result in a 2-3 vol % liquid yield loss with a conventional catalyst, the loss in selectivity with the new platinum–rhenium catalyst is less than 1 vol % and is only about 2% after a 100°F temperature rise.

In normal nonregenerative or semiregenerative operations, yield losses are allowed to increase until it is economic to regenerate or replace the catalyst. Normally, yield losses are not allowed to increase beyond 2 vol %. In view of the improved yield stability of the platinum–rhenium catalysts it is now possible to operate at much lower yield losses or to operate to much lower activity levels than has been common in the past. The ability to increase temperature rise during the process cycle to high levels should be considered in the design of new plants using this catalyst.

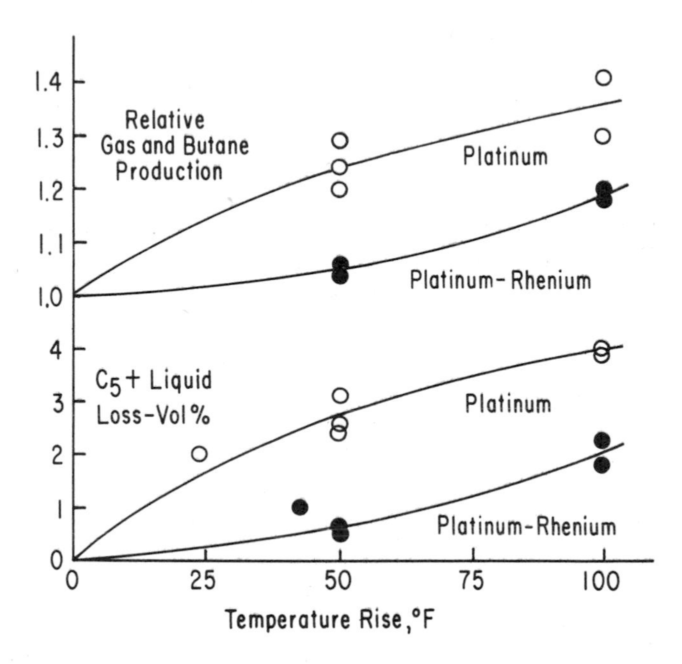

Figure 4. Catalyst selectivity loss. Pressure, 100–500 psig; research octane, 95–100.

Another outstanding characteristic of the platinum–rhenium catalyst is its low aging rate with respect to activity. Typical comparisons with conventional catalysts are shown in Figure 5 where operating conditions were essentially constant for the two types of catalysts. The comparisons obtained to date indicate that the catalyst life for the platinum–rhenium catalyst when operated to a given activity decline or temperature rise is about four times that of the conventional catalysts. Under these conditions the yield loss from the rhenium containing catalyst is considerably less than from the platinum catalyst. If the platinum–rhenium catalyst were allowed to operate to the same yield loss as the rhenium-free catalyst, the life would be considerably greater than the four times indicated above.

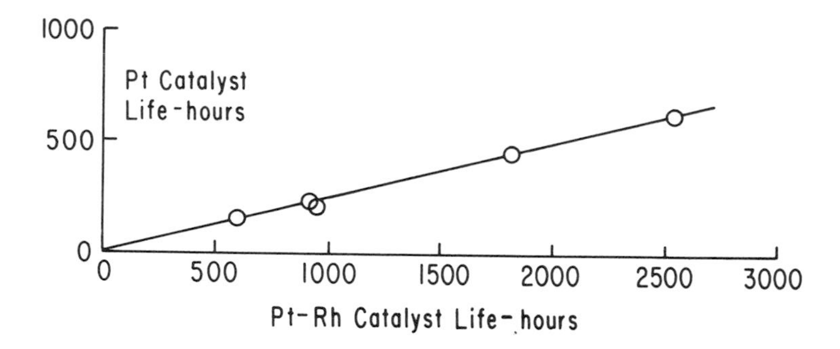

Figure 5. Relative catalyst life (equivalent activity basis)

Regeneration

Regeneration requirements of the catalytic reforming process depend on the severity of the operation. Processes, such as Ultraforming and Powerforming, are designed for low pressure, low recycle gas rate, high space velocity, and high octane levels. Frequent regeneration of the catalyst is required. Most other reforming processes have operated at higher pressure, higher recycle gas rate, and lower space velocity, conditions being selected so that catalyst regeneration is required only after extended operation—at least 4 months. Regeneration of platinum catalysts has required the development of special techniques, each of the many reforming processes having proprietary regeneration conditions.

Data on the regeneration characteristics of the platinum–rhenium catalyst are obviously not as extensive as on the conventional platinum catalysts. Some difficulties have been reported. However, the best performance of the platinum–rhenium catalyst requires modification of the normal process, regeneration, and rejuveniation conditions. Furthermore, these catalysts have been regenerated successfully both in laboratory and commercial unit equipment.

Additives and Impurities

In the catalytic reforming of naphthas there are a number of nonhydrocarbon materials which play an important part in the performance of the catalyst. Sulfur is a poison for the reforming catalyst. There appears to be evidence developing that the platinum–rhenium catalysts may be more sensitive to sulfur than the conventional catalysts. Effective pretreatment of the feed stock to maintain sulfur at low levels is desirable.

Water concentration is important in catalyst performance. The catalyst bases are gel structures, and some water is needed to avoid dehydration of the gels. On the other hand, excessive water vapor will promote loss of surface area and activity of the catalyst. Halide addition is important in the maintenance of catalyst activity, and proper control of water vapor and halide is needed to prolong catalyst life.

Magnaforming

In addition to the announcement by Chevron of the improved platinum–rhenium catalyst, Englehard and Atlantic Richfield announced the Magnaforming process as an improved modification of the conventional catalytic reforming processing technique. This process decreases the amount of recycle gas to the first reforming reactor and increases the recycle gas to the last reactor. In the primary reforming reactor, the dehydrogenation of naphthenes is rapid and highly endothermic. There is a large temperature drop with the result that the system approaches the thermodynamic equilibrium of the naphthene–aromatic–hydrogen system. This slows down the dehydrogenation reaction relative to the hydrocracking reaction. By decreasing hydrogen to the first reactor this effect is minimized, and gas production caused by hydrocracking is decreased.

Also in Magnaforming, a four- rather than three-reactor system is recommended, there being an ascending inlet temperature from the first to the last reactor. The recycle gas to feed mole ratio might be 2.5 to 3 in the first reactor and 9 to 10 in the last reactor. Liquid yield advantages of a least 1 to 2% and increased hydrogen production have been obtained with both platinum and platinum–rhenium catalysts. This increased yield reportedly justifies a 6.5% increase in investment for the Magnaforming design over conventional three-reactor systems.

Future Reformer Design

The process design requirements of future catalytic reformers will depend considerably on the degree to which the oil refining industry will attempt to maintain the octane quality of lead free gasoline. Since catalytic reforming will be the high octane refining process, it is doubtful if any future design will be for less than 100 research octane number clear. Present day Ultraformers and Powerformers will certainly produce this quality product at high liquid yields. It is believed, however, that the introduction of the platinum–rhenium catalyst will allow the design of lower investment semiregenerative type units at high octanes, low operating pressure, and high liquid recoveries. The economics of decreasing the operating pressure in a semiregenerative unit from the conventional 350 psig pressure to 200 psig is detailed in Table XI. Operation was at 100 octane with a Mid-Continent type naphtha. The lower pres-

Table XI. Economics of Low Pressure Reforming, Semiregenerative Process for a Mid-Contient Naphtha

Pressure, psig		*200*	*350*
Feed, BPSD		*10,000*	
Product yields		*Table X*	
Investment, $		3,800,000	3,400,000
Incremental Costs			200 lb minus 350 lb
Investment, $			+400,000
Feed and Products, $/day			
Ext. butanes		$2.05/Bbl	+472
Hydrogen		20¢/Mcuft	−580
Dry gas FOE		$1.90/Bbl	+328
10 lb gasoline		$5.35/Bbl	−1605
Direct costs, $/day			+352
Indirect costs, $/day			+109
Total New Credits			
$/day			924
$/year			304,920
Investment payout, yrs			1.3

Table XII. Catalytic Reforming of *n*-Heptane Using a Platinum Catalyst

Temperature, °F	*925*			
w/hr/w	*2.0*			
Pressure, psig	*100*	*200*	*350*	*500*
C_7 conversion, mole %	98	97	98	97
Selectivity, mole %				
Hydrocracking	39	54	77	88
Dehydrocyclization	61	46	23	12

sure case had a 2 vol % yield advantage in C_5+ liquid, which increased to 3 vol % on a 10 lb RVP basis. This yield advantage, after adjusting for other credits and debits, paid out the incremental investment in 1.3 years before taxes.

Future Catalyst Development

The future demand for high octane gasoline production by catalytic reforming requires extensive conversion of paraffins to aromatics. It is

Table XIII. Catalytic Reforming with Theoretical Catalyst Improvement

Feed	*56 API MC naphtha*		
Pressure, psig	*360*		
Research clear	*95*		
Hydrocracking reduction	Base	50%	100%
Hydrogen, CF/bbl	950	1300	1650
Dry gas, wt %	8.1	4.1	—
Butanes, vol %	6.9	3.4	—
C_5 + liquid, vol %	80.7	85.9	91.0
10lb RVP liquid, vol %	91.3	98.5	105.8
Extraneous butanes, vol %	3.7	9.2	14.7

readily apparent from the product distribution obtained in the reforming of naphtha that considerable by-product hydrocracking occurs in this conversion process. This is illustrated more clearly in the work of Hettinger *et al.* (*6*) on the reactions of *n*-heptane over platinum–alumina catalyst. Typical data are presented in Table XII. At 500 psig selectivity was 88% for hydrocracking and only 12% for dehydrocyclization. Pressure is an effective processing tool for controlling these two reactions, but even at 200 psig pressure the selectivity was more for hydrocracking than dehydrocyclization. The conversion of higher boiling paraffins is more selective than shown for *n*-heptane. However, an improvement in catalyst formulation is needed to decrease the hydrocracking reaction.

The theoretical effect of an improvement in catalyst performance on the product distribution from reforming a 56° API Mid-Continent heavy naphtha at 350 psig and 95 research octane number clear is shown in Table XIII. Liquid product yields could be increased from 81 to 86% if hydrocracking were decreased 50%. The perfect catalyst would give a yield of 91%. A decrease in hydrocracking would give a lower vapor pressure product, and the resultant effect on the 10 lb RVP gasoline yield would be more marked. An improved catalyst would also increase the production of valuable by-product hydrogen.

There is considerable room for improvement in formulating the catalytic reforming catalyst to avoid the high liquid losses when reforming to high octane levels. A major breakthrough in this area is needed to minimize the high costs involved in producing high octane lead free gasoline.

Literature Cited

(1) Jacobson, R. L., Kluksdahl, H. E., McCoy, C. S., Davis, R. W., *Am. Petrol. Inst., Div. Refining, 34th Midyear Meetg.* (May 13, 1969).

(2) Blue, E. M., *Hydrocarbon Processing* (1969) **48** (9), 141.
(3) Kopf, F. W., Decker, W. H., Pfefferle, W. C., Dalson, M. H., Nevison, J. A., *Hydrocarbon Processing* (1969) **48** (5), 111.
(4) Bonner & Moore Associates, Inc., American Petroleum Institute, "U.S. Motor Gasoline Economics," Vol. 1, June 1967.
(5) Hettinger, W. P., Jr., Keith, C. D., Gring, J. L., Teter, J. W., *Ind. Eng. Chem.* (1955) **47**, 719.

RECEIVED May 25, 1970.

6

The Continuing Development of Hydrocracking

J. W. SCOTT and A. G. BRIDGE

Chevron Research Co., P.O. Box 1627, Richmond, Calif. 94802

During the past decade hydrocracking has developed into a major refining process. Commercial feedstocks now range from naphthas to residua. The adaptation of the process to such a broad scope of applications has required intensive and essentially continuous process and catalyst research. In this paper the catalytic requirements throughout this range of application are examined. An attempt is made to generalize the relationship between performance of commercial units and laboratory kinetic measurements. The important effects of diffusion within catalyst particles are examined. Some of the catalytic challenges for hydrocracking during the next decade are discussed.

Hydrocracking has seen rapid growth during the 1960's (*1, 2, 3*). When the modern version of hydrocracking was announced in 1959 (*4*), capacity was 1000 barrels per day. As the decade ended, capacity onstream or under construction was approaching 1,000,000 barrels per day. In the latter part of the decade much of the new hydrocracking technology proved applicable to solving the important problem of fuel oil desulfurization. The transfer of this technology permitted rapid accommodation to the need for desulfurization of high molecular weight petroleum fractions. Consequently, the world's petroleum industry was able to commit to the construction of nearly 700,000 barrels per day of fuel oil desulfurization within a few years. As indicated in Figure 1, the total worldwide capacity for these two related processes is approaching 1,700,000 barrels per day.

This review considers the present state of hydrocracking technology. Experience gained in its extension to fuel oil desulfurization is drawn upon where appropriate. Laboratory results and theoretical calculations are compared with commercial practice, and, refining needs for the next

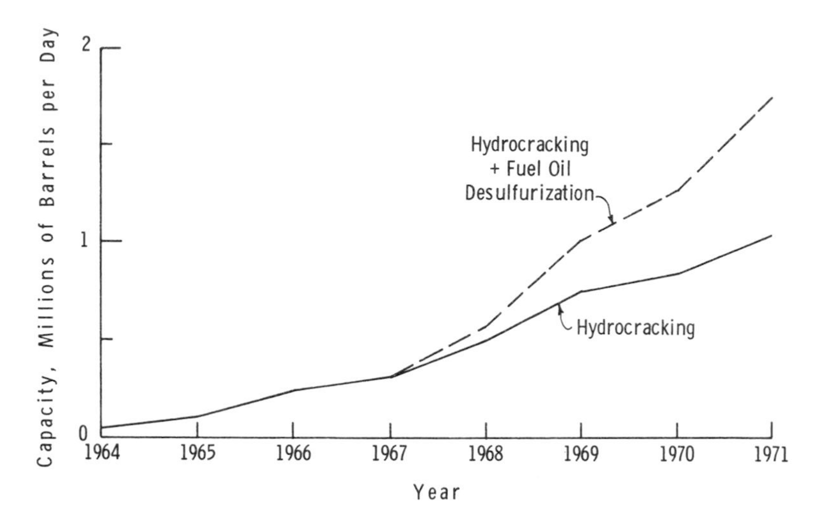

Figure 1. Worldwide growth of hydrocracking and fuel oil desulfurization

decade are discussed. The effect these needs will likely have upon the development of hydrocracking and related technologies are considered.

The Hydrocracking Process

Hydrocracking is probably the most versatile of modern petroleum processes. This versatility has been achieved by the development of specific families of catalysts, of processing schemes designed to allow these catalysts to function efficiently, and of optimum refining relationships between hydrocracking and other refining processes.

The choice of processing schemes for a given hydrocracking application depends on the feedstock to be processed and the desired product yield structure and quality. Figures 2 and 3 show two of the more common flow arrangements in simplified schematic form. The two-stage plant (Figure 2) was developed largely to convert cracked and straight run gas oils into gasoline at high yields. In the first stage the feed is hydroprocessed to remove basic impurities like nitrogen-containing hydrocarbons. The second stage hydrocracks this product at extinction recycle. The recycle still cut point is selected to maximize a desired product. As indicated on the figure, distillation facilities may be placed either between the stages or after the second stage.

The single-stage plant (Figure 3) can be used to produce gasoline but is most advantageously used to produce middle distillate from heavy vacuum gas oils. The operation is again extinction recycle above a cut point which can be above 700°F. These two flow schemes represent cases in which substantial cracking is desired. In other applications, hydrocracking is less important than the hydroconversion of nonhydrocarbon constituents, like sulfur, nitrogen, oxygen, and metal-containing molecules. The first stage of a two-stage plant usually falls into this category as do the various processes used in fuel oil desulfurization.

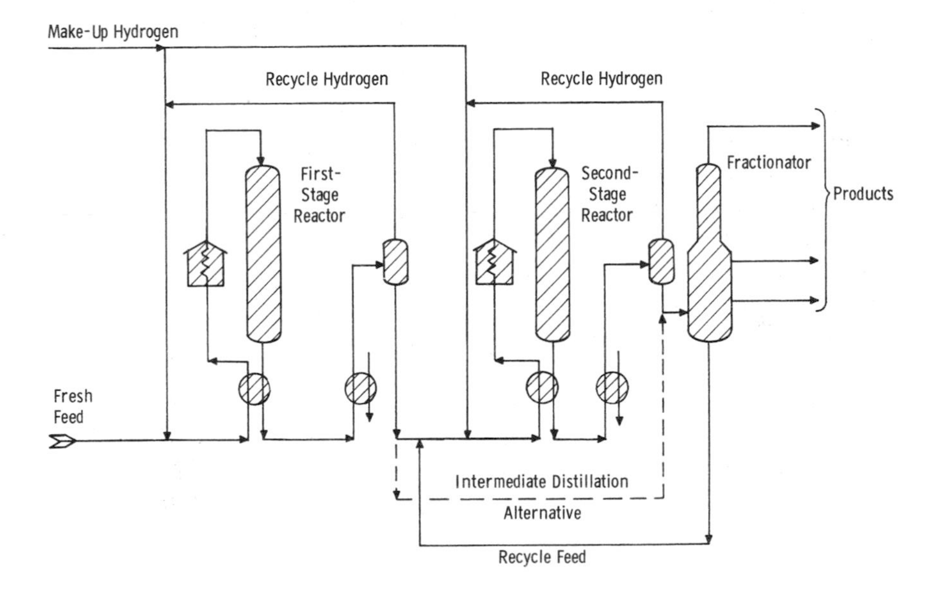

Figure 2. Simplified hydrocracking two-stage process flow diagram

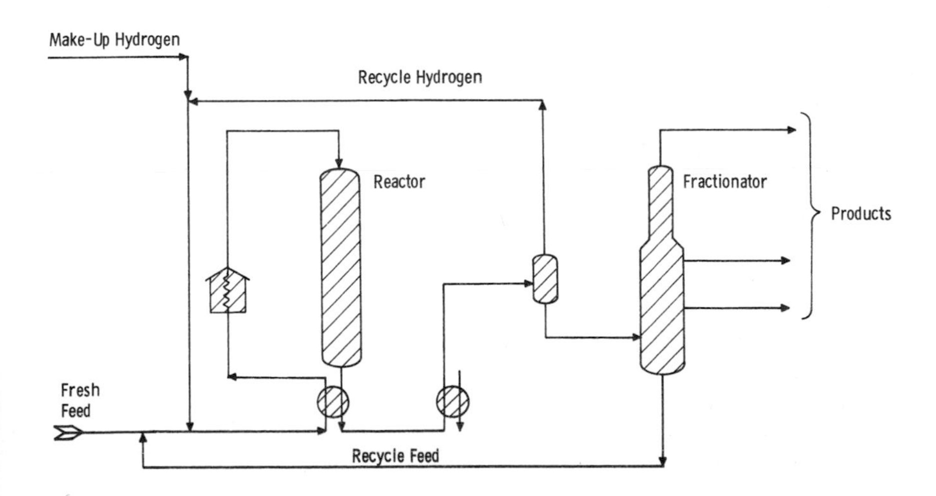

Figure 3. Simplified hydrocracking single-stage process flow diagram

The Catalysts

Hydrocracking catalysts are dual functional, containing both hydrogenation and cracking sites. The best choice of catalyst for a specific situation requires a particular balance between these two functions. Table I shows some applications of hydrocracking and the characteristics of the families of catalysts that have evolved for accomplishing the indicated conversions.

When hydrocracking to LPG and gasoline, strong cracking activity is required; this is achieved by using strongly acidic materials, which include both amorphous silica-aluminas and crystalline aluminosilicates. The acidity of these materials promotes reactions which lead to high iso–normal ratios in the light paraffin products, low methane and ethane production, and conservation of monocyclic rings. The hydrogenation component acts to reduce the concentration of coke precursors and maintains the effectiveness of the cracking sites. Catalysts can then be operated for long times at economic processing conditions.

When hydrocracking gas oils to jet fuel and middle distillate, catalysts with less acidity and stronger hydrogenation activities are used. This type of catalyst is also valuable in producing high viscosity index lubricating oils and in the general area of hydrocracking residual fractions such as solvent-deasphalted oils and residua.

Table I. Hydrocracking Catalyst Types

	Catalyst Characteristics			
Desired Reaction	*Acidity*	*Hydrogenation Activity*	*Surface Area*	*Porosity*
Hydrocracking Conversion				
A. Naphthas to LPG (*2, 5*) Gas oils to gasoline (*6, 7*)	strong	moderate	high	low to moderate
B. Gas oils to jet and middle distillate (*1, 8, 9*)	moderate	strong	high	moderate to high
Gas oils to high v.i. lubricating oils (*10, 11*)				
Solvent deasphalted oils and residua to lighter products (*12, 13*)				
Hydroconversion of Non-hydrocarbon Constituents				
Sulfur and nitrogen in gas oils (*14, 15*)	weak	strong	moderate	high
Sulfur and metals in residua (*16, 17*)				

For hydroconversion of nonhydrocarbon constituents, catalysts with weak acidity are used where cracking is undesirable. Strong hydrogenation activity is needed, particularly with heavy feedstocks containing high molecular weight aromatics.

Another important catalyst characteristic is porosity. Particularly when heavy feeds are processed, high pore volumes and pore diameters are required to reduce pore diffusion limitations. These limitations occur when the intrinsic rate of reaction is high compared with the rate of diffusion of the reactants through the catalyst particle to the active surface. The catalyst is then not used effectively, and reaction rates and selectivity become functions of particle size. If the kinetics of the reaction are known, it is possible to estimate from theory the reaction rate or threshold above which a catalyst of known size will begin to exhibit diffusion limitations.

Referring again to Figure 3, certain generalizations can be made about catalyst porosity. For hydrocracking to LPG and gasoline, pore diffusion effects are usually absent. High surface areas (above 300 square meters/gram) and low to moderate porosity are used (from 12 A pore diameter with crystalline acidic components to 50 A or more with amorphous materials). Moderately acidic catalysts tend to have a more open structure (from 50 to perhaps 100 A pore diameter). With conversion of high molecular weight nonhydrocarbon constituents, pore diffusion can exert a large influence. Examples are given later illustrating the need for greater porosities (above 80 A pore diameter) for catalysts used in these processes.

Despite the complexity of hydroconversion processes, reaction kinetics can often be expressed in simple terms. Figure 4 shows, for example, the first-order nature of the hydrodenitrification reaction. The data were obtained when a heavy California coker distillate was processed in the pilot plant over a weakly acidic catalyst containing both a Group VI and a Group VIII hydrogenation component. First-order behavior describes the data over a range of product nitrogen covering four orders of magnitude.

Residuum desulfurization kinetics are generally not first order. Figure 5 illustrates this with a first-order plot for desulfurization of Arabian light residuum. On this type of plot a first-order reaction would yield a straight line with a slope corresponding to the reaction rate constant. The over-all desulfurization reaction is not therefore first order and can in fact be represented by second-order kinetics. However, the figure shows that it may also be considered as the sum of two competing first-order reactions. The rates of desulfurization of the oil and asphaltene fractions are reasonably well represented as first-order reactions whose

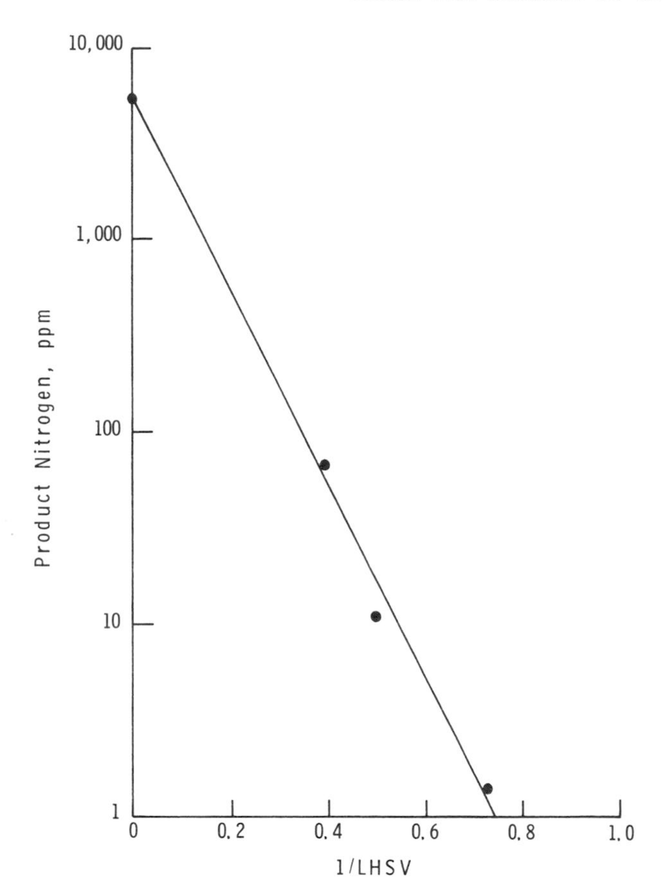

Figure 4. Hydrodenitrification kinetics for a California coker gas oil

sum is apparently second order. Thus, if used with discretion, first-order techniques may be used to analyze many hydroconversion reaction systems.

Basic constituents in a feed or product can affect the acid component of hydrocracking catalysts. Catalysts which utilize the strongly acidic crystalline aluminosilicates are somewhat less sensitive to ammonia and basic nitrogen compounds than are those containing amorphous acidic components. In the presence of nitrogen, catalyst temperatures must be increased to maintain conversion until an equilibrium is reached in the competition between reacting hydrocarbons and basic compounds such as ammonia. This is illustrated in Figure 6 for California straight run gas oils denitrified to different levels. Here, catalyst activity loss (°F) accompanying equilibration to a constant cracking temperature is plotted against feed nitrogen content. To allow these strongly acid catalysts to operate at low temperatures, feedstocks are generally denitrified in a separate stage.

The important effect of catalyst acid strength on product distribution is illustrated in Table II. Typical product yields are shown when an Arabian straight run vacuum gas oil is hydrocracked over a moderately acidic catalyst in a single-stage plant and a strongly acidic catalyst in a two-stage plant. In the single-stage operation, the feed is hydrocracked efficiently to jet fuel with minimum formation of lighter products. The mild acidity of the catalyst is effective in producing low freeze point jet fuel and, when run at higher cut points, low pour point synthetic diesel from high pour point feed stocks. Operation with a strongly acidic catalyst in the second stage of a two-stage plant is described by yields at recycle cut points of 550° and 430°F. At the higher cut point, product yields are similar to the single-stage plant. However, because the balance between hydrogenation and cracking has been changed, over-all hydrogen consumption is lower, jet smoke point is lower, and naphtha octane numbers are higher. As the cut point is reduced further, high yields of

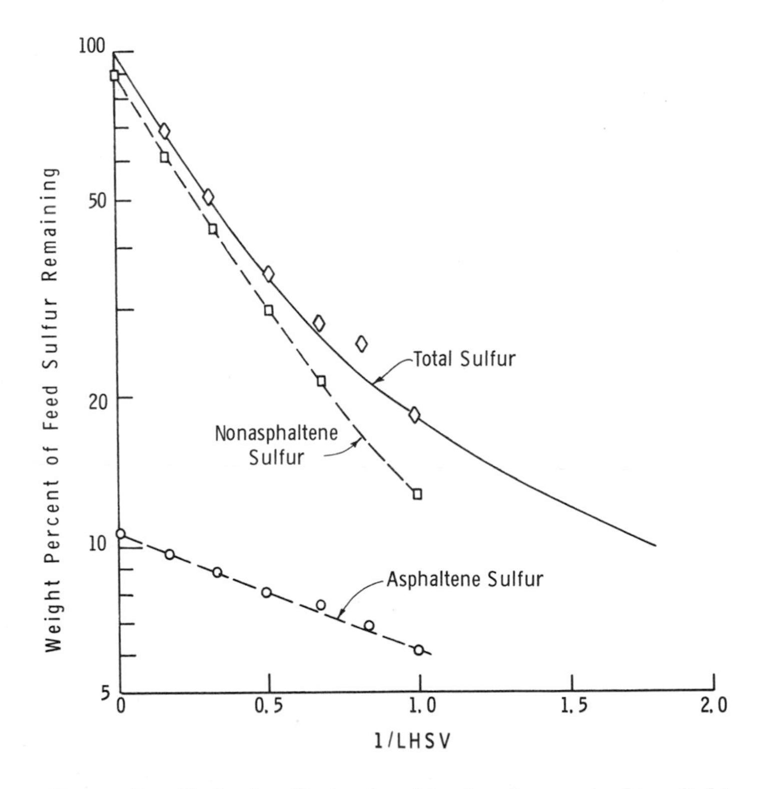

Figure 5. Hydrodesulfurization kinetics for an Arabian light atmospheric residuum

Table II. Effect of Catalyst Acid Strength on Product Distribution

Feedstock: Arabian Vacuum Gas Oil

Catalyst Type	*Moderately Acidic*	*Strongly Acidic*	
Recycle Cut Point, °F	*550*	*550*	*430*
Product Yields			
C_1–C_3, wt %	3.0	1.2	3.2
Butanes, LV %	9.1	8.0	16.2
C_5–180°F, LV %	15.1	14.5	32.2
180°–280°F, LV % or 180°F +, LV %	22.5	24.6	73.3
280°–550°F, LV %	68.8	69.1	
C_5 +, LV %	106.4	108.2	105.5
H_2 consumption, SCF/B	1350	1240	1525

high octane number light naphtha are produced. The heavy naphtha is an ideal catalytic reformer feedstock.

To produce the maximum yield of finished gasoline in a hydrocracker–reformer combination, the hydrocracker should be operated to give maximum liquid yields, followed by a modern low pressure catalytic reformer to give the desired octane improvement (*6*). More severe hydrocracker operation produces higher octane naphtha but leads to an increase in the production of butanes with reduced yields of finished

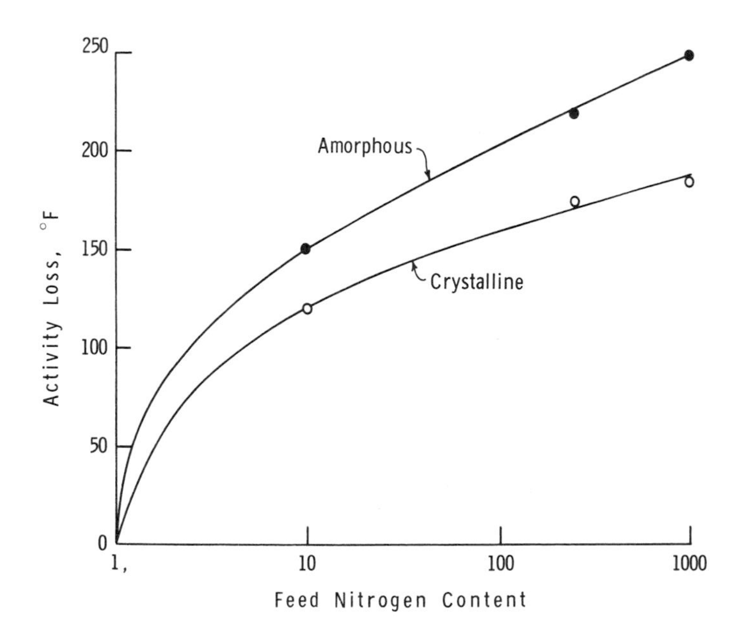

Figure 6. Effect of feed nitrogen content on strongly acidic hydrocracking catalyst activity (amorphous and crystalline components)

gasoline. This type of operation is used commercially when butanes or LPG are desired products.

Feedstock Effects

During the last decade, there has been a definite trend towards heavier hydrocracker feedstocks (2). Recognition of the interactions between feed molecular weight and catalyst chemical and physical properties has been necessary to support this trend.

Figure 7 shows the effect of feed molecular weight on the reaction rates observed with strongly acidic hydrocracking catalysts. These data were obtained with an early version of an amorphous catalyst. They may, however, be used to illustrate general trends involving feed character and molecular weight.

First-order reaction rates normalized to a constant temperature and pressure are shown for a variety of pure hydrocarbons. For this display, a line is drawn connecting the points for *n*-paraffins. Other points are displayed for isoparaffins, naphthenes, aromatics, and polycyclics.

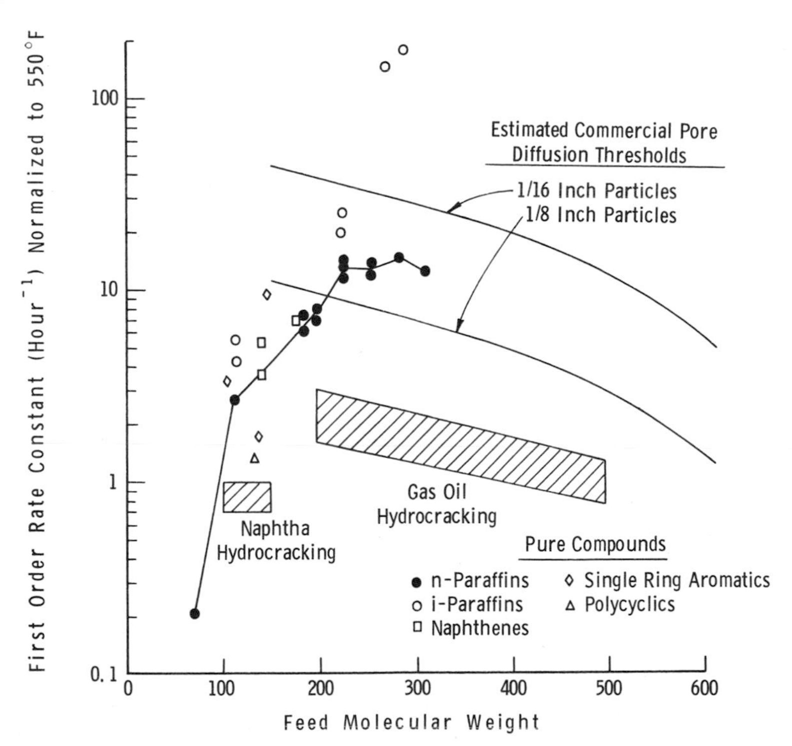

Figure 7. Laboratory and commercial reaction rates using strongly catalysts

The pure compound rate constants were measured with 20–28 mesh catalyst particles and reflect intrinsic rates (—*i.e.*, rates free from diffusion effects). Estimated pore diffusion thresholds are shown for 1/8-inch and 1/16-inch catalyst sizes. These curves show the approximate reaction rate constants above which pore diffusion effects may be observed for these two catalyst sizes. These thresholds were calculated using pore diffusion theory for first-order reactions (*18*). Effective diffusivities were estimated using the Wilke-Chang correlation (*19*) and applying a tortuosity of 4.0. The pure compound data were obtained by G. E. Langlois and co-workers in our laboratories. Product yields and suggested reaction mechanisms for hydrocracking many of these compounds have been published elsewhere (*20–25*).

The pure compound cracking rates may be compared with typical reaction rates that were found commercially with wide-boiling petroleum fractions. Commercial naphtha hydrocracking data agree reasonably closely. Gas oil hydrocracking rates are lower and decrease with feedstock molecular weight. This is probably caused by the heavy aromatic molecules' inhibiting the acid function of the catalyst. Despite this suppression of reaction rates, careful balancing of hydrogenation and cracking functions produces catalysts which operate efficiently at economical processing conditions. Particle size effects have not been observed with these catalysts, and this is consistent with the position of the commercial data relative to the diffusion limit curves.

Figure 8 shows the effect of molecular weight on hydroconversion rate constants observed with typical catalysts of lower acidity and higher hydrogenation activity. The hydrocracking of residuum is nearly 10 times more difficult than gas oil hydrocracking. This is attributed to the large asphaltenic molecules present in the residua. The residuum conversion rate constants shown in Figure 8 represent data for straight run residua containing a wide range of molecular sizes. As indicated in Figure 5, if the heavy asphaltenic molecules are processed by themselves, much lower reaction rates are observed. Solvent deasphalted oils are correspondingly easier to process than straight run residua. The reaction rate constants for denitrification of gas oils, and desulfurization and demetalation of residua, are substantially higher than the hydrocracking rate constants. These nonhydrocarbon constituents therefore can be removed selectively with minimum hydrocracking of the parent molecule.

Two calculated pore diffusion threshold curves are shown in Figure 8. For gas oil hydrocracking, the observed reaction rate constants are not high enough to lead to problems; this is supported by commercial hydrocracking experience. The high denitrification rate constants suggest that pore diffusion problems could occur with active catalysts at high

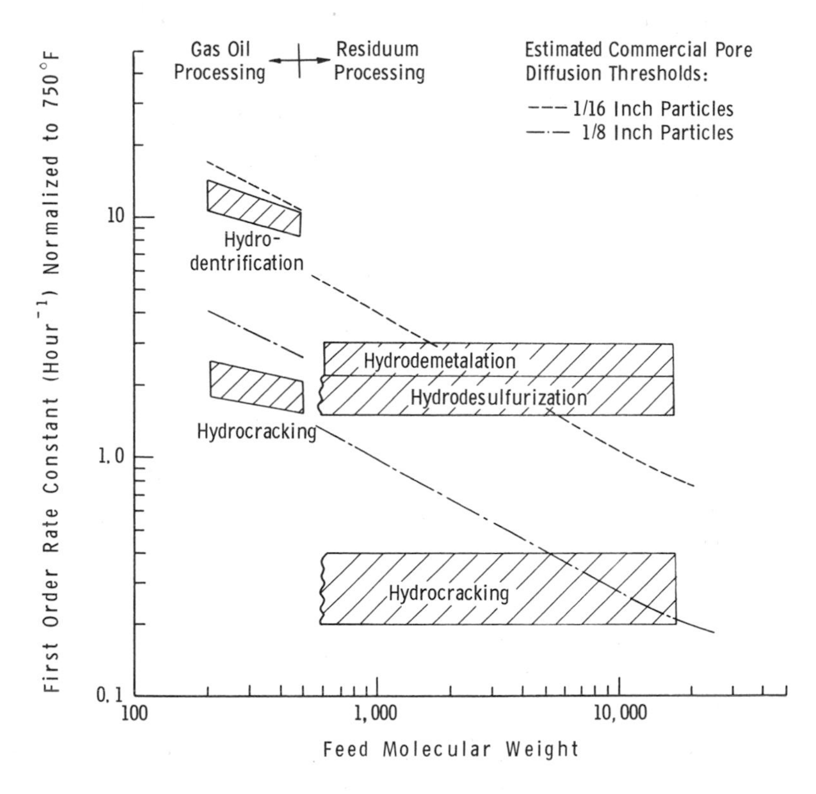

Figure 8. Effect of feed molecular weight on hydroconversion rates using moderately and weakly acidic catalysts

temperatures. The estimated diffusion limits for residuum processing with 1/16-inch catalysts suggest that demetalation would be influenced markedly and desulfurization to a lesser extent.

An effect of pore diffusion in residuum demetallation is illustrated in Figure 9, which shows nickel and vanadium concentration profiles measured through a catalyst pill after residuum desulfurizing service. The catalyst originally contained neither of these metals. These profiles confirm that the rate of reaction of the metal-containing molecules in the feed (particularly the vanadium compounds) is high compared with their rate of diffusion.

Besides influencing over-all reaction rates, pore diffusion can cause changes in selectivity. An extreme example of this was observed (*26*) when a high molecular weight California solvent-deasphalted oil was hydrocracked over a small pore size palladium zeolite catalyst at high temperatures. The feedstock gravity was 16.4° API, and 70% boiled above 966°F. The resulting product distribution is compared with that

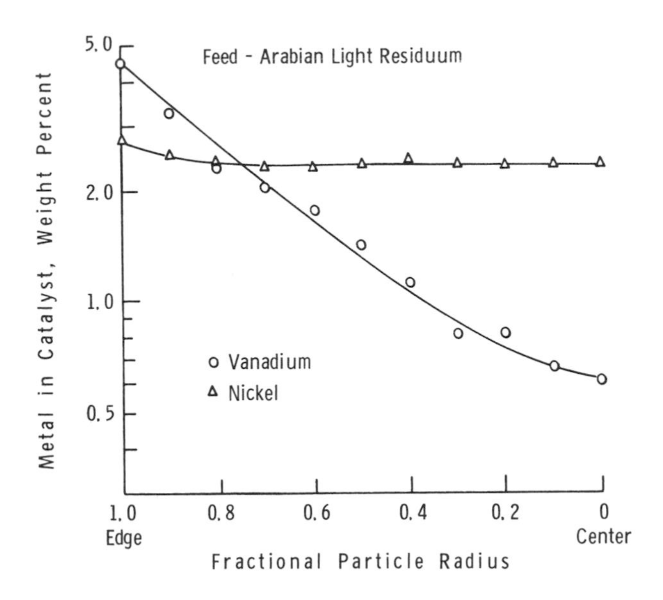

Figure 9. Desulfurization catalyst analysis after pilot plant test

of a larger pore amorphous hydrocracking catalyst in Figure 10. At the lower temperatures, the more conventional catalyst was considerably more active for cracking the heavy feed. As the temperature was raised, this activity relationship was maintained; however, the selectivity of the catalysts was profoundly different.

At 850°F., the zeolite produced a combined yield of propane and butane of 46 wt % compared with only 6 wt % with the large pore catalyst. With the zeolite very little synthetic product boiling above 400°F was found. Methane and ethane production was minimal with both catalysts. Selectivity differences of this type arise when molecular dimensions approach or exceed catalyst pore size. The large oil molecules can crack only on acid sites near the mouths of the pores or in the macropores. The small number of sites so located requires that high temperatures be used to initiate appreciable cracking. The resulting cracked products, however, can diffuse into the micropores where they encounter many acid sites at the high temperature to give the ultimate products of strong acid cracking: C_3's, C_4's, and C_5's.

Although the effect of pore diffusion on catalyst activity is usually undesirable, its effect on selectivity can sometimes be used to advantage, as reported by Weisz and others at the Mobil Research Laboratories (*27*).

Diffusion Limits and the Choice of Reactor Systems

From the foregoing dicussion it is apparent that residuum hydroconversion processes can be influenced adversely by pore diffusion limitations. Increasing the catalyst porosity can alleviate the problem although increased porosity is usually accompanied by a decrease in total catalytic surface area. Decreasing the catalyst particle size would ultimately eliminate the problem. However, a different type of reaction system would be required since the conventional fixed bed would experience excessive pressure drops if very fine particles were used. A fluidized system using small particles does not suffer from this limitation. However, staging of the fluidized reaction system is required to minimize the harmful effects that backmixing can have on reaction efficiency and selectivity.

Figure 11 shows estimates of the effect that decreasing catalyst particle size would have on over-all reactor size in a diffusion-limited situation. Calculations were carried out for a first-order reaction at 80% conversion, constant catalyst temperature, and constant reactor throughput. Three different diffusion-limited cases were chosen in which 1/8-inch catalyst had effectiveness factors of 0.6, 0.3, and 0.1. Effectiveness factors were defined in the usual way (*24*). A decreasing effectiveness factor from 0.6 to 0.1 could be caused by considering feeds of increasing molecular weight, shown directionally on the figure. Plug flow fixed

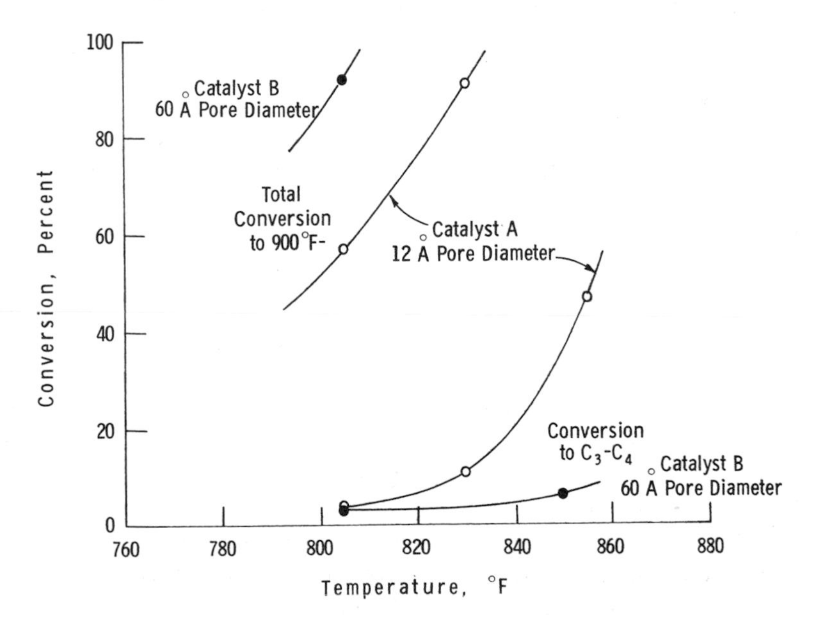

Figure 10. Interaction between catalyst pore size and feed molecular weight in hydrocracking a California solvent deasphalted oil

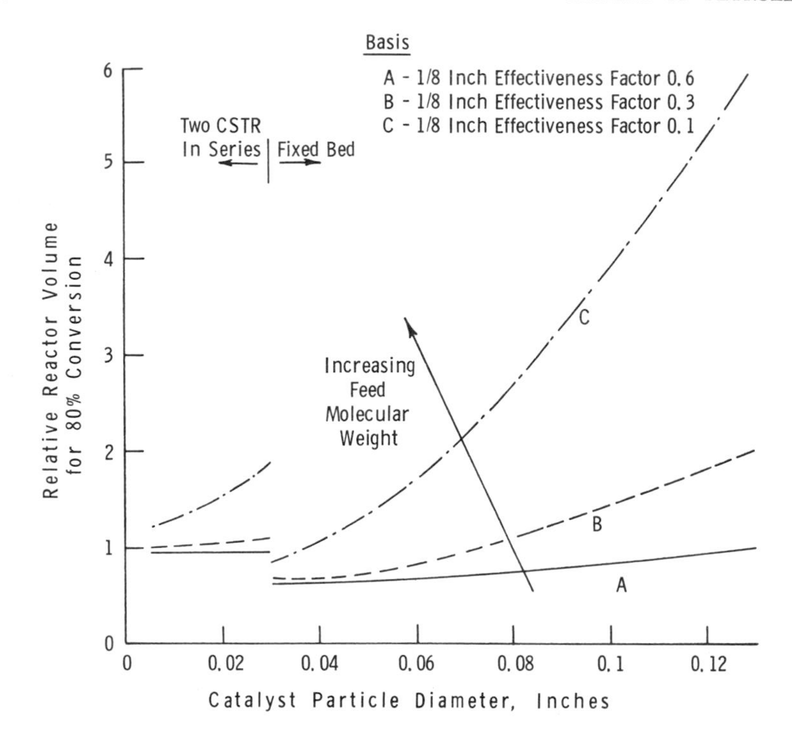

Figure 11. Effect of catalyst size and effectiveness factor in different reaction systems

bed processing was assumed for particle sizes from 1/8 to 1/32 inch. Fluid bed processing (equivalent to two continuous stirred tank reactors in series) was assumed for particles smaller than 1/32 inch. Fifteen percent extra reactor volume was added in the fluid bed cases to allow for bed expansion.

As expected, the fluid bed system looks most attractive when the reaction is severely diffusion limited. Here the required reactor volume is only 20–30% of that required by a fixed bed of 1/8-inch particles. However, reducing the particle size in a fixed bed from 1/8 to 1/16 inch would accomplish a similar reduction. As a result, we can conclude that from a reaction kinetics viewpoint a reaction must be limited severely by pore diffusion before the extra reactor volume required for small particles in fluidized beds is offset by their increased activity. Few residuum reactions are presently hindered to this extent.

Adlington and Thompson (*28*) have pointed out that the advantages of a fluidized system are not likely to be associated with any marked improvement in catalyst efficiency. Other factors, like freedom from pressure drop increases when processing feeds which contain particulate matter, or ability to add and withdraw catalyst during operation, are more likely to influence the choice.

Summary and a Consideration of Refining Needs of the Next Decade

Application of the modern hydrocracking process began with conversion of refractory light gas oils to gasoline. Catalytic and process developments permitted a rapid expansion of process applications both to lighter and to heavier feedstocks. Research on variations of catalyst function and morphology resulted in the evolution of families of catalysts, each with a particular application and selectivity. Intensive development led to industrial catalysts which possess combinations of acidity, hydrogenation activity, surface area, and porosity suitable for a wide range of applications. Incidental to these developments, effective catalysts became available for the removal of nonhydrocarbon constituents—sulfur, nitrogen, oxygen, and metal-containing molecules.

There are some differences evident between laboratory and commercial achievements, but these now are predominantly in the area of heavy feed processing. Reaction systems employing fixed catalyst beds are used extensively although staged fluidized beds of catalyst are used with some high molecular weight feedstocks (*2*). That hydrocracking is the economic choice in many refinery applications is evident from the rapid growth of capacity during the 1960's.

The next decade will see both economic and environmental factors affecting process selection and operation. Within the refineries, there is a need to manage effluent streams carefully to reduce the release of pollutants. At the same time, there is a need to remove from products the precursors of pollutants that would otherwise be released during intended use. These requirements combine to place a double burden on the refineries.

We can expect continuing growth of hydrocracking because it is able to satisfy both of these requirements. Pollutants are removed as ammonia and hydrogen sulfide. The technology is readily available (*29*) for recovering these materials for commercial use. Water requirements are moderate, and the self-contained nature of a hydrocracker makes it particularly attractive when close environmental control is required.

Some major trends and some uncertainties are evident in the markets for gasoline, jet fuel, and fuel oil. Production of unleaded gasoline will have a profound effect upon refinery processes and their interrelationship. Hydrocracking, combined with low pressure catalytic reforming (*30*), appears particularly appropriate for the production of unleaded gasoline. Hydrocracking catalyst selectivity can be adjusted to maximize production of prime reformer feed. The products are clean, olefin free, and the refiner can trim octane numbers readily because of the flexibility of the reforming process.

Fortunately, the production of high quality jet fuel is entirely consistent with the optimum combination of hydrocracking and reforming. We can expect continuing evolution of hydrocracking catalysts which can maximize jet fuel at the expense of gasoline and maximize reformer feed at the expense of light gasoline components.

The 1970's will see substantial reduction in the amount of higher sulfur residua moving into domestic fuel markets. If natural gas fuel supplies are adequate, these residua will be converted to more valuable light products. The development of hydroprocessing catalysts has proceeded to the point where it is certain that desulfurization, and perhaps hydrocracking, will be practiced on higher quality residua before the decade is over.

Lastly, we may expect to see more consideration given to the all-hydrocracking refinery. This type of refinery offers the advantages of simplicity, flexibility for producing gasoline, jet, and lubricating oils, and favorable environmental factors both within the refinery and in the marketplace.

The hydrocracking technology and the families of industrial hydrocracking catalysts developed during the past decade have excellent prospects for increased application during the next.

Literature Cited

(1) Scott, J. W., Robbers, J. A., Mason, H. F., Paterson, N. J., Kozlowski, R. H., "Isomax: A New Hydrocracking Process in Large-Scale Commercial Use," *World Petrol. Congr., Frankfurt* (1963).
(2) Scott, J. W., Paterson, N. J., "Advances in Hydrocracking," *World Petrol. Congr., Mexico City* (1967).
(3) Bachman, W. A., "Forecast for the Seventies," *Oil Gas J.* (1969) **67** (45), 154.
(4) Stormont, D. H., "New Process Has Big Possibilities," *Oil Gas J.* (1965) **63** (52), 130.
(5) *Oil Gas J.* (1965) **63** (52), 130.
(6) Kittrell, J. R., Langlois, G. E., Scott, J. W., "Optimizing Hydrocracking-Reforming Brings Profit," *Oil Gas J.* (1969) **67** (20), 118.
(7) "Unicracking—JHC Satisfies Processing Requirements," *Oil Gas J.* (1969) **67** (24), 76.
(8) Kozlowski, R. H., Mason, H. F., Scott, J. W., "Preparation of Low Freezing Jet Fuels by Isocracking," *Ind. Eng. Chem., Process Design Develop.* (1962) **1** (4), 276.
(9) Watkins, C. H., Jacobs, W. L., "How to Get High Quality Jet Fuels," *Oil Gas J.* (1969) **67** (47), 94.
(10) *Oil Gas J., Newsletter* (1969) **67** (20), 1.
(11) Beuther, H., Donaldson, R. E., Henke, A. M., "Hydrotreating to Produce High Viscosity Index Lubricating Oils," *Ind. Eng. Chem., Prod. Res. Develop.* (1964) **3** (3), 174.

(12) Stormont, D. H., Hoot, C., "World's Biggest Hydrocracker," *Oil Gas J.* (1966) **64** (17), 145.
(13) Blue, E. M., Harvey, P. D., Lance, R. P., Rossi, W. J., "Operation of the Richmond Isomax Complex," *API, Div. Ref., 33rd Midyear Meeting, Philadelphia* (May 1968).
(14) Flinn, R. A., Larson, O. A., Beuther, H., "How Easy Is Hydrodenitrogenation," *Hydrocarbon Processing Petrol. Refiner* (1963) **42** (9), 129.
(15) van Deemter, J. J., "Trickle Hydrodesulfurization—Case History," *European Symp. Chem. Reaction Eng., 3rd, Amsterdam* (Sept. 1964).
(16) Scott, J. W., Bridge, A. G., Christensen, R. I., Gould, G. D., "The Chevron RDS Isomax Process," Japan Petroleum Institute, *Fuel Oil Desulfurization Symp., Tokyo* (March 1970).
(17) Beuther, H., Schmid, B. K., "Reaction Mechanisms and Rates in Residue Hydrodesulfurization," *World Petrol. Congr., Mexico City* (1967).
(18) Satterfield, C. N., "Mass Transfer in Heterogeneous Catalysis," MIT Press, Cambridge, 1970.
(19) Wilke, C. R., Chang, P., "Correlation of Diffusion Coefficients in Dilute Solutions," *Am. Inst. Chem. Engrs. J.* (1955) **1,** 270.
(20) Langlois, G. E., Sullivan, R. F., "Chemistry of Hydrocracking," ADVAN. CHEM. SER. (1970) **97,** 38.
(21) Egan, C. J., Langlois, G. E., White, R. J., "Selective Hydrocracking of C_9 to C_{12} Alkylcyclohexanes on Acidic Catalysts. Evidence for the Paring Reaction," *J. Am. Chem. Soc.* (1962) **84,** 1204.
(22) Langlois, G. E., Sullivan, R. F., Egan, C. J., "The Effect of Sulfiding a Nickel on Silica-Alumina Catalyst," *J. Phys. Chem.* (1966) **70,** 3666.
(23) Sullivan, R. F., Egan, C. J., Langlois, G. E., "Hydrocracking of Alkylbenzenes and Polycyclic Aromatic Hydrocarbons on Acidic Catalysts. Evidence for Cyclization of the Side Chains," *J. Catalysis* (1964) **3,** 183.
(24) Sullivan, R. F., Egan, C. J., Langlois, G. E., Sieg, R. P., "A New Reaction that Occurs in the Hydrocracking of Certain Aromatic Hydrocarbons," *J. Am. Chem. Soc.* (1961) **83,** 1156.
(25) White, R. J., Egan, C. J., Langlois, G. E., "Reactions of Large Cycloalkane Rings in Hydrocracking," *J. Phys. Chem.* (1964) **68,** 3085.
(26) "Production of Light Hyrocarbon Gases by Hydrocracking High Boiling Hydrocarbons," Chevron Research Co., U. S. Patent **3,385,782** (May 28, 1968).
(27) Chen, N. Y., Weisz, P. B., "Molecular Engineering of Shape Selective Catalysts," *Chem. Eng. Progr., Symp. Ser.* (1967) **73** (63), 86.
(28) Adlington, D., Thompson, E., "Desulfurization in Fixed and Fluidized Bed Catalyst Systems," *European Chem. Reactor Technol. Symp., Amsterdam* (Sept. 1964).
(29) "Ammonia Recovery Process," Chevron Research Co., U. S. Patent **3,335,071** (Aug. 8, 1967).
(30) Jacobson, R. L., Kluksdahl, H. E., McCoy, C. S., Davis, R. W., "Platinum-Rhenium Catalysts: A Major New Catalytic Reforming Development," *API, Div. Ref., Midyear Meetg., 34th,* Chicago (May 1969).

RECEIVED June 5, 1970.

7

Alkylation and Isomerization

W. L. LAFFERTY, JR. and R. W. STOKELD

Texaco Inc., Beacon, N. Y.

The histories of alkylation, a widely accepted commercial process, and isomerization, a process with fewer commercial applications, are traced from their early stages of development to their present position in the petroleum refining industry. The factors which have affected the growth of these processes are reviewed along with recent advances in technology and developments in processing equipment which have improved their status in the industry. The various types of alkylation and isomerization processes based on different catalyst systems and equipment configurations are compared. The general effects of operating variables, charge stock compositions, and charge stock purification systems are also discussed. The future roles of alkylation and isomerization in petroleum refining are considered with respect to the continued growth of the gasoline market and the possible changes in gasoline specifications which may be required for automotive emission control.

In petroleum refining the term alkylation generally applies to the catalytic reaction of isobutane with various light olefins to produce highly branched paraffins for use in high octane gasoline. Alkylation processes have attained exceptional importance in the petroleum industry over the past 30 years. The rapid growth of alkylation created large demands for isobutane, and to help supply these demands, butane isomerization processes were developed. The isomerization technology was later applied to higher molecular weight paraffins, and as a result, various isomerization processes were developed to complement alkylation as a tool for gasoline manufacture. This paper reviews the development history of the alkylation and isomerization processes, considers their present status in the industry, and predicts their future role in the manufacture of petroleum products.

Alkylation

Development and Growth. Alkylation became commercially feasible when thermal and catalytic cracking processes began to be used widely in the 1930's. In cracking the heavy components of crude oil to supply the growing demand for gasoline, substantial quantities of light hydrocarbons, such as ethylene, propylene, and butylenes, and isobutane, were produced. Only limited quantities of these light hydrocarbons could be used in gasoline because of their high vapor pressure. Consequently, these compounds had to be used largely for fuel gas, but their availability set chemists and engineers to working on the classical task of finding more profitable uses for the by-products from the cracking processes. Consequently, alkylation was developed (*1*).

Most of the petroleum companies were experimenting with the alkylation reaction following the discovery of the reaction by Ipatieff and co-workers using an aluminum chloride catalyst. Texaco Inc., Shell Development in the Hague, Netherlands, and the Anglo Iranian Oil Co., Ltd. in England each experimented independently with a sulfuric acid catalyst at about the same time. Following a publication by Anglo Iranian (*2*) in 1938, Humble Oil and Refining Co., with the cooperation of Anglo Iranian, put in a commercial plant at Baytown, Tex. by converting and adapting a polymerization unit to alkylation. Texaco Inc. designed a completely new plant and put in a commercial plant at Port Arthur, Tex.

Five companies—Anglo-Iranian Oil Co., Ltd., Humble Oil and Refining Co., Shell Development Co., Standard Oil Development Co., and Texaco Inc. issued a joint report and introduced the sulfuric acid alkylation process to the industry at the Chicago meeting of the Petroleum Institute, Nov. 17, 1939. Other companies participating in some of the early development work on alkylation processes were Union Oil Co., Stratford Engineering Corp., The M. W. Kellogg Co., Universal Oil Products, and Phillips Petroleum Co. (*2, 3*). By the end of 1939, six commercial sulfuric acid alkylation units were on stream making 3525 BPD of aviation alkylate. Eight more units were under construction or contracted for, which would make an additional 9000 BPD of alkylate (*3*).

The early experimental work showed that the alkylation reactions could be carried out thermally and by using hydrofluoric acid or aluminum chloride as catalysts. Successful processes were developed with the hydrofluoric acid catalyst, and the first commercial HF unit started up on Christmas Day 1942, having a capacity of 1950 BPD (*3*). Other commercial HF alkylation units followed quickly. The thermal and aluminum chloride-catalyzed alkylation processes had limited commercial acceptance.

The excellent characteristics of alkylate were quickly recognized. It had a high octane rating, a high heat of combustion per pound, a low vapor pressure, and a desirable boiling range (*1*, *3*). In addition, alkylate had good lead susceptibility with low or, in some cases, negative sensitivity (RON-MON). Thus, alkylate was found to be extremely well suited for use in aviation fuel or motor gasoline blending.

World War II caused a tremendous increase in the demand for fuels, and the ability of the petroleum industry to gear up and supply these demands played an important part in the outcome of the war (*3*). The alkylation processes provided the basic ingredient of the huge quantities of aviation fuel which were used by the military forces, and, as shown in Figure 1, alkylation capacity increased rapidly through the war years. Between 1939 and 1946, 59 alkylation units were built. By the end of the war capacity had reached 169,000 PBD (*5*).

When World War II ended, the demand for aviation alkylate dropped, and refiners had excess alkylate which they started using in motor gasoline. About half of the alkylation units were soon shut down, and some of the units were dismantled because the octane levels of motor gasoline were not high enough to justify the continued use of this relatively expensive process (*6*).

The Korean War caused the demand for aviation alkylate to jump again in 1950, but only about 35 alkylation units were in operation at the

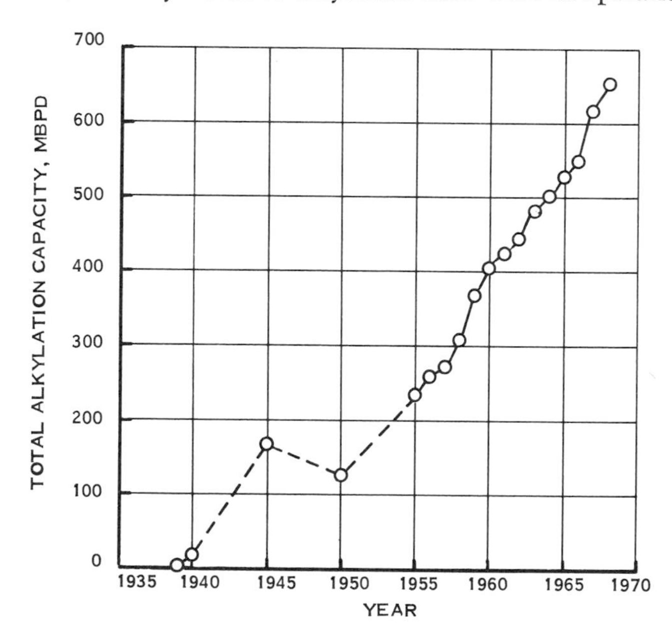

Figure 1. Growth in alkylation capacity

time (*4*). In compliance with government requests, refiners stopped using alkylate in motor gasoline and started expanding their aviation alkylate production. Government contract guarantees and allowances for fast tax write-offs provided the incentives for installing new capacity. By the time the Korean War ended, the motor gasoline octane race was underway, and the alkylate used for aviation fuel during the war began to be used as a major high octane motor fuel component (*7*).

The demand for aviation gasoline started to decline in the middle and late 1950's because of the increasing use of jet aircraft by the military forces and commercial airlines (*8, 9*). However, the steady increase in motor gasoline octane requirements and the increasing volume of gasoline required by the growing automobile population provided an alternate use for large quantities of alkylate so that there was never a serious excess of capacity. In fact, the alkylation capacity increased through the late 1950's, and the growth trend has continued to the present time.

Types of Alkylation Processes. Almost all of the commercial alkylation units have been based on catalytic processes using H_2SO_4 or HF. The H_2SO_4 process was commercialized first, and it has led the industry throughout the history of alkylation as shown in Figure 2 (*5, 8, 10*).

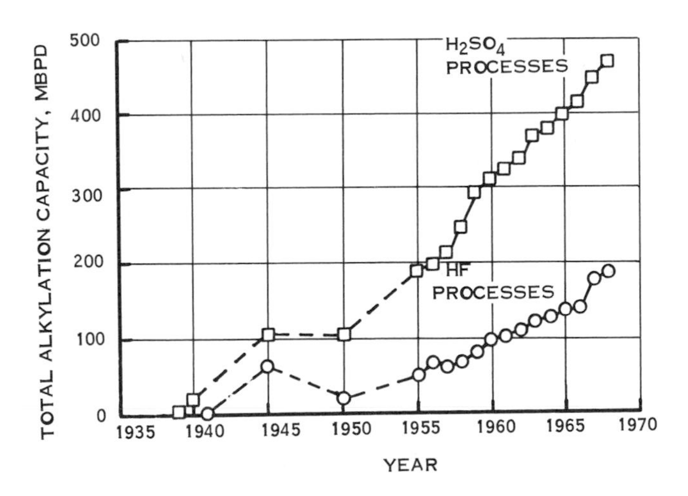

Figure 2. Comparison of H_2SO_4 and HF growth

About twice as many H_2SO_4 units as HF units have been installed. In 1968, H_2SO_4 alkylation processes accounted for about 71% of the total capacity.

The ranges of operating conditions normally used in commercial H_2SO_4 and HF alkylation processes are shown in Table I. Both processes operate in the liquid phase at relatively low temperatures. For the best

Table I. Ranges of Operating Conditions

Process Catalyst	H_2SO_4	*HF*
Temperature, °F	35–60	60–125
Isobutane/olefin feed	3–12	3–12
Olefin space velocity, vo/hr/vc	0.1–0.6	—
Olefin Contact Time, min	20–30	8–20
Catalyst acidity, wt %	88–95	80–95
Acid in emulsion, vol %	40–60	25–80

results, H_2SO_4 units must operate in the range of 35°–55°F when charging butylenes. When processing olefin feed mixtures containing high percentages of propylene, the preferred temperature is in the range of 50°–60°F. Higher temperatures can be used in HF alkylation. The HF unit operating temperature is usually set by available cooling water temperatures although higher octanes can be produced at temperatures around 60°F. Pressure does not have a significant effect on the alkylation reactions as long as it is set high enough to maintain the liquid phase. The isobutane-to-olefin ratio has a very pronounced effect on the alkylation reactions; increasing the volume ratio up to about 12/1 improves alkylate yields and octanes in both processes. Isobutane-to-olefin ratios of 5/1 or higher are generally used in commercial operations. In H_2SO_4 alkylation, the olefin space velocity is generally in the range of 0.1–0.6 volumes of olefin per hour per volume of acid in the contactor. Olefin contact times (based on the empty reactor volume) of 8–20 minutes have been used in HF designs. Too short a contact time or too high a space velocity can adversely affect alkylate yields and quality whereas designing for longer contact times or lower space velocities than required results in an unnecessarily large reactor. H_2SO_4 units normally operate with an acid strength in the range of 88–95 wt %. HF acid strengths may be between 80 and 95 wt %. Acid feed rates to the reactor are set to give 40–60 vol % acid in the emulsion for H_2SO_4 alkylation and 25–80 vol % for HF alkylation. More detailed discussions of the effects of operating variables may be found in the literature (*1, 11, 12, 13, 14*).

There has always been considerable controversy between the proponents of H_2SO_4 alkylation processes and those of HF alkylation processes over which process is superior. A comparison of published yield and quality ranges for H_2SO_4 and HF alkylation of propylene, butylene, and pentylene is presented in Table II (*4*). The advocates of the H_2SO_4 process point out that it produces higher octanes, utilizes isobutane more efficiently owing to less self-alkylation, requires a less expensive acid, and uses a safer system with the higher boiling point acid. [Self-alkylation is the reaction of two isobutane molecules and an olefin molecule. The two isobutane molecules combine to form isooctane and the olefin is saturated by hydrogen transfer to produce a light paraffin. This reaction is undesirable since it increases isobutane consumption and the olefin reacts to form a compound too light to be used in gasoline.] The HF alkylation advocates claim lower investment and operating costs, lower acid consumption, and the ability to operate at cooling water temperatures without refrigeration being required. Owing to the high cost of HF, a recovery system has always been built into an HF alkylation unit. Until recently this was not the case with H_2SO_4 units. The acid consumption data in Table II are for the H_2SO_4 process without recovery and for the HF process after recovery of the catalyst.

The arguments for both processes probably are valid for certain cases so there is no way to settle this controversy once and for all. However, the record indicates that the majority of petroleum refiners prefer the H_2SO_4 system.

As mentioned previously, the use of aluminum chloride alkylation has been very limited in the petroleum refining industry. The aluminum chloride catalyst, being somewhat more difficult to handle and regenerate, could not compete economically with H_2SO_4 and HF catalysts for propylene and butylene alkylation. However, aluminum chloride will catalyze ethylene alkylation whereas H_2SO_4 and HF will not. In the past, ethylene alkylation has not been used much because of the higher olefin feed cost (*15*).

Thermal alkylation was never a totally successful commercial process because of the severe operating conditions required. The reaction was carried out in a heater coil with temperatures of 900°–975°F, pressures in the range of 3000–5000 psig, and contact times of 2–7 seconds (*16*). Polymerization of the olefins occurred readily under these conditions, and low olefin concentrations had to be used to minimize undesirable side reactions. Ethylene could be alkylated more readily than the higher molecular weight olefins, and either normal butane or isobutane could react with the olefin. In general, the yields and quality of the product were not equal to those obtained with catalytic alkylation.

Advances in Alkylation Technology. Significant improvements have been made in the alkylation process during its 30 years commercial lifetime. Table III summarizes the major technical and mechanical advances.

Reactor design improvements are one of the most important developments. The early plants used a pump and time–tank reactor system which was designed to mix the reactants intimately with the catalyst and to remove the exothermic heat of reaction for temperature control (*15*). As commercial experience was accumulated, the importance of creating a high interfacial area between the acid and hydrocarbon phases by providing a high degree of turbulence and shear in the reactor system was recognized. Good mixing promoted higher concentrations of dissolved

Table II. Quality and

Catalyst	H_2SO_4		
Olefin Feed	$C_3^=$	$C_4^=$	$C_5^=$
Yields, vol % Olefin			
Propane	—	—	—
Butane	—	—	—
Pentanes	4–8	4–12	10–30
Alkylate	175–180	160–175	165–185
Isobutane Consumption, vol % olefin	115–135	100–125	90–125
Acid Consumption, lb/gal alkylate	1.0–3.0	0.4–1.2	0.7–2.0
Alkylate Octane,			
RON clear	87–91	91–97	88–92
MON clear	85–88	88–95	86–90

isobutane in the acid phase which were required for the desirable reactions to proceed with the exclusion of the undesirable reactions. Since the early reactors were inadequate in this respect, new reactor designs evolved which improved the degree of acid–hydrocarbon contacting.

The criticality of good temperature control was also recognized as commercial experience was gained. Controlling the temperature of the reaction mixture in the preferred range was absolutely essential for good alkylation. Inadequate temperature control resulted in decreased alkylate yields and octanes and increased acid consumption. Therefore, to avoid these penalties the new reactor designs included improved temperature control techniques as well as improved mixing (*12*).

The two most commonly used reactor systems which grew out of the reactor development work for H_2SO_4 alkylation are the Stratford Engineering Co.'s Stratco contactor and the M. W. Kellogg Corp. cascade reactor. These reactors are designed to provide good mixing through optimum use of power input with minimum operating costs. Efficient temperature control is provided in the Stratco contactor by flashing a portion of the reactor effluent through a heat exchanger in the reaction zone. In the Kellogg cascade reactor, refrigeration for temperature control is provided directly by flashing hydrocarbons in the reactor. More complete descriptions of these reactor systems and some of their other advantages are in the literature (*4, 12, 17*).

HF alkylation reactors have undergone more drastic changes than H_2SO_4 reactors. In the middle 1950's a Stratco contactor similar to that for H_2SO_4 alkylation was used. Later a vertical reactor was developed in which the reactants were bubbled up through liquid HF. Cooling was

Yield Data for Alkylation

HF		
$C_3^=$	$C_4^=$	$C_5^=$
7–20	—	—
0–5	0–3	—
	3–15	15–40
175–180	160–173	165–175
120–170	110–130	90–125
0.011	0.009	0.011
87–91	90–96	87–91
85–89	88–94	85–89

provided in this reactor by a U-tube exchanger. In some of the more recent designs, a shell and tube heat exchanger is used for the reactor. The reactants flow through the shell side at high enough velocities to give the turbulence needed for mixing, and cooling water flows through the tubes for temperature control (*13*).

Control of the important operating parameters other than temperature and the degree of mixing has also led to significant improvements in alkylation (*17*). For example, the effects of isobutane–olefin ratio, acid strength, acid concentration in the reaction mixture, and olefin space velocity have been recognized, and efforts to control these variables in the preferred ranges have resulted in more profitable operations.

Improvements in feed preparation and pretreatment have made important contributions to the advances in alkylation technology (*12, 17*). The ability to design better fractionators has made higher quality feedstocks available, and feed pretreatment facilities have been developed to remove water, mercaptans, sulfides, and diolefins effectively. The benefits of these advances have been realized as higher alkylate yields and octanes, lower acid consumption, and reduced corrosion.

Product treatment has also been improved (*17*). Bauxite treating, hot water washing, and electrostatic precipitation are some of the significant developments which have improved product quality and reduced fouling and corrosion in downstream equipment.

The sulfuric acid recovery process (SARP), developed jointly by Texaco Inc. and Stratford Engineering Corp. to reduce the acid consumption in H_2SO_4 alkylation units, was another contribution to alkylation technology (*18*). In this process the spent acid from an alkylation

Table III. Advances in Alkylation Technology

Improved reactors
(A) better mixing
(B) better temperature control
Recognition and control of operating variables
Improved feed preparation
Improved product treatment
Sulfuric acid recovery process
Catalyst promoters
Mechanical and construction improvements

unit reacts with a portion of the olefin feed to form dialkyl sulfates. The dialkyl sulfates are extracted from the reaction mixture with isobutane, and the extract is charged to the alkylation unit. Thus, SARP recovers H_2SO_4 from the spent acid and returns it to the alkylation unit while rejecting the water and oil diluents. The effect is a net reduction in acid consumption. Two commercial SARP units have been installed.

Various catalyst promoters have been reported to improve alkylation operations (*9*, *19*). These additives are claimed to increase alkylate yields, raise octanes, lower isobutane requirements, and reduce acid consumption by promoting the desired reactions. Plant trials have been made to test these catalyst promoters, but there is no indication in the literature that such additives have been widely used.

Alkylation, like other refinery processes, has benefited greatly from improvements in mechanical equipment and construction techniques. For example, the use of centrifugal compressors and a better understanding of the metallurgy and type of equipment required for handling strong acids have significantly reduced the operating and maintenance costs on modern alkylation units (*17*, *20*).

Alkylation Today. Today, alkylation is the key tool in petroleum refineries for controlling motor fuel mid-boiling range quality and MON levels. The steady succession of developments and improvements in alkylation technology has resulted in a process which is capable of producing 91–97 RON clear gasoline (89–95 MON) in extremely good yields with only a fraction of the acid consumption that was required when the process was first developed. Alkylate end points have declined through the years so that they are now in the range of 350°–400°F, and rerunning is no longer necessary (*17*).

As shown in Table II, the alkylation of propylene and pentylenes results in higher acid consumptions and lower product octanes than butylene alkylation. Because of these penalties, only butylenes were alkylated for many years. However, the rapidly increasing demand for alkylate as a motor gasoline component has forced refiners to include propylene and, to a lesser extent, pentylenes in their alkylation feeds.

Considerable effort has been put into minimizing the adverse effects of these olefins. It was found that alkylating propylene and pentylenes in a mixture with butylenes promoted the desired reactions and reduced the octane and acid consumption penalties. Furthermore, by optimizing temperature, isobutane-to-olefin ratio, acid strength, and other variables, the deleterious effects of propylene and pentylenes in the feed can be minimized (*4, 8, 21*). The decision as to how much of these olefins to include in the alkylation unit feed depends on many different factors, such as their value relative to alkylate, butylene and isobutane avails, alkylate volume and octane requirements, acid costs, etc.

Refiners have a choice of using either alkylation or polymerization to synthesize gasoline from lower molecular weight hydrocarbons. Almost without exception in the last two decades, alkylation has been chosen because it produces a superior gasoline product. Alkylate does not contain olefins which require gum inhibitors or hydrogenation to improve stability, and it has a higher RON and MON than polymer gasoline.

Isomerization

Advances in Isomerization Technology. To appreciate the development of isomerization technology the equilibrium relationships between the various isomers must be considered. As shown in Figure 3 the equilibrium values for the more highly branched isomers are generally favored by lower temperatures. This is readily apparent for butanes and pentanes. For hexane paraffins lower temperatures favor the formation of 2,2-dimethylbutane (92.3 RON, clear) whereas temperature has a much smaller effect on 2,3-dimethylbutane (103.5 RON, clear). Within a temperature range of 200°–500°F the 2,3-dimethylbutane equilibrium concentration remains between 8.5 and 9.5%. The equilibrium ratio for methylcyclopentane and cyclohexane is included in these curves since it is very costly to separate them from light straight run naphthas.

The significance of the equilibrium relationships become more apparent to the refiner when the unleaded research and motor octane values for each carbon group are volumetrically blended and plotted *vs.* temperature. Such a curve is shown in Figure 4. The sensitivity for the C_5 and C_6 paraffins is about 1–2 numbers on a clear basis *vs.* 10–13 for the C_6 naphthenes. All of the octane numbers for these components are shown in Table IV.

Isomerization catalysts were developed along two paths—by Friedel-Crafts halide systems or by dual site heterogeneous catalysts, originating with the commercial introduction of platinum-aluminas for catalytic reforming in the 1940's. The Friedel-Crafts systems (aluminum chloride–hydrocarbon complexes) were used exclusively during the early stages of

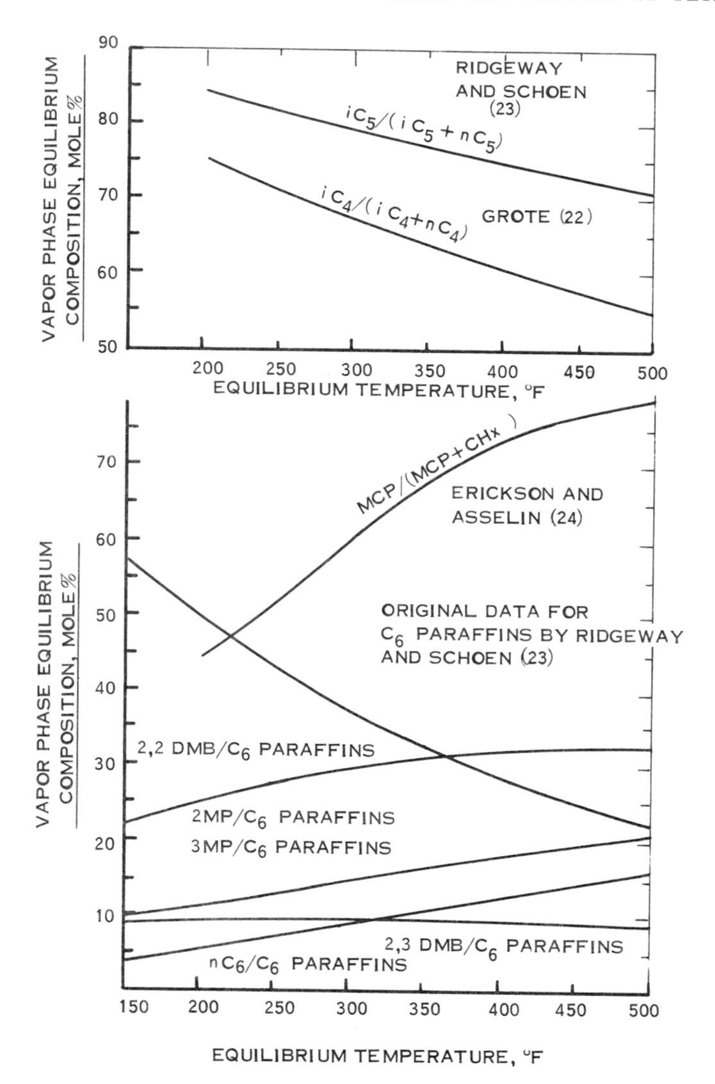

Figure 3. Vapor-phase equilibrium composition for butanes, pentanes, and hexanes

World War II when the first commercial isomerization processes were introduced to manufacture isobutane as a feedstock for aviation alkylate. The Friedel-Crafts catalysts are highly active at 100°–200°F whereas the conventional platinum-alumina reforming catalysts give reasonable reaction rates only above 850°F.

Friedel-Crafts Catalyst Systems. The isomerization activity of the Friedel-Crafts catalysts has not been excelled—they will produce near equilibrium conversions in the 100°–200°F temperature range. However, owing to their extreme reactivity they tend to react indiscriminately with reactants or products destroying themselves unless the feedstock is purified properly, operating conditions are carefully controlled, and suppressors are used to reduce the number of side reactions. This is particularly true as the carbon number of the isomerization medium is increased from a C_4 through a C_6 (*25*). Recent commercial developments have been designed to minimize these side reactions by (1) introducing compounds such as antimony trichloride as suppressors or (2) by adding hydrogen to the reaction complex when isomerizing C_5/C_6 paraffins (*26*).

DUAL FUNCTIONAL CATALYST SYSTEMS. Lawrance and Rawlings have given an excellent review of dual functional catalysts (*27*). Table V is an adaptation of one of their illustrations showing how changes in the catalyst base will enable the reaction temperature to be dropped from 850°F or above to about 500°F. However, the real breakthrough has come by making the acid function of the dual site catalysts more acidic. This has been done by exposing the platinum alumina base to either aluminum chloride or selected organic chlorides at 400°–900°F (*28, 29, 30, 31*). Such catalysts are highly active in the 200°–400°F temperature region. This is only a slightly higher temperature region than the 100°–200°F

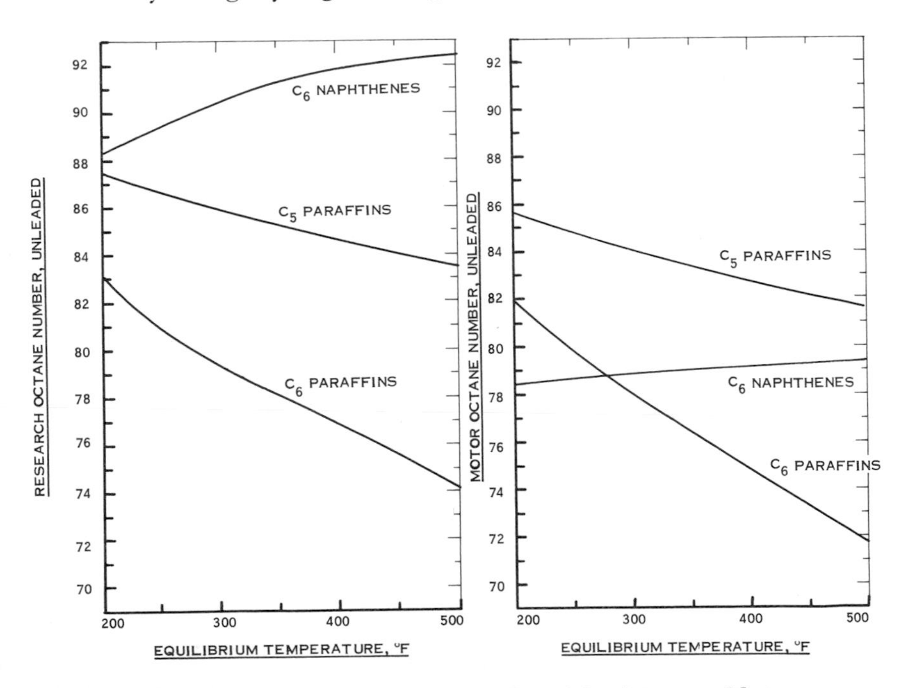

Figure 4. Research and motor octane numbers (clear) for equilibrium conversions for C_5 and C_6 isomerization. Motor octane number curves estimated from data presented by Erickson et al. *(24). Basic information contained in Figure 3 and Table IV.*

Table IV. Isomer

	Boiling Point, °F	*Vapor Pressure At 100°F*
Isobutane	10.9	72.2
n–Butane	31.1	51.6
Isopentane	82.1	20.4
n–Pentane	96.9	15.6
Cyclopentane	120.7	9.9
2,2–Dimethylbutane	121.5	9.9
2,3–Dimethylbutane	136.4	7.4
2–Methylpentane	140.5	6.8
3–Methylpentane	145.9	6.1
n–Hexane	155.7	5.0
Methylcyclopentane	161.3	4.5
Cyclohexane	177.3	3.3

[a] Sources of octane numbers: Phillips Petrolum Co. Bulletin No. 521 (1962) for butanes, all others obtained from Erickson and Asselin (*24*). Much of the octane numbers in the latter case represent blending values established by extensive UOP testing.

Table V. Development of Dual-Functional Isomerization Catalyst Systems[a] (*27*)

Platinum-alumina
(T = 850°–950°F)
↓
Platinum-alumina-silica
(T = 700°–850°F)
↓
Platinum-alumina-boria
(T = 600°–700°F)
↓
Platinum Y-type zeolite
(T = 600°–650°F)
↓
Platinum-mordenite
(T = 500°–550°F)

Reactions of platinum-alumina with aluminum chloride
(T = 200–400°F)
↓
Reactions of platinum-alumina with organic chlorides
(T = 200°–400°F)

[a] In many cases the metal hydrogenation component may be either platinum or palladium.

Properties[a]

RON, Clear	*MON, Clear*	*RON, 3cc TEL*	*MON, 3cc TEL*
102.1	97.0	118.3	—
94.0	89.1	104.1	104.7
92.6	90.3	103.5	106.9
61.7	61.3	84.7	83.6
101.3	85.0	111.1	95.2
92.3	92.9	103.4	114.6
103.5	94.3	112.0	109.7
73.4	72.9	92.2	92.4
74.5	74.0	92.3	92.6
34.0	25.0	65.3	63.5
95.0	80.0	104.3	91.1
83.0	77.2	97.4	87.3

reaction temperatures used for Friedel-Crafts catalyst systems. Although this higher temperature is unfavorable from equilibrium considerations, the effect of the temperature differences is minimized since isomerization selectivity for the dual function catalysts is extremely high. For instance, only small amounts of C_1–C_4 paraffins are found as by-products in C_5/C_6 *n*-paraffin isomerization.

Both the Friedel-Crafts and the dual functional catalyst systems require careful charge stock purification to prevent water, organic oxygenates, and sulfur compounds from reacting with the chlorine components making up these catalysts. Should any water enter the charge system, serious catalyst deactivation and corrosion problems will ensue. For the dual functional catalysts, extremely cautious catalyst loading procedures must also be exercised unless the catalyst may be activated or regenerated *in situ*.

Types of Isomerization Processes. A number of isomerization processes have been written up in the technical (*24, 26, 27, 32, 33, 34, 35, 36*) or the patent (*27, 28, 29, 30, 31, 37, 38, 39*) literature. All but three (*24, 26, 27*) appear to use dual functional catalyst systems. Further remarks in this section are limited mainly to the three processes which have been commercialized according to the literature.

Salient features of the three commercialized processes are shown in Table VI. The Shell Development Co's liquid phase isomerization process uses an improved Friedel-Crafts catalyst system consisting of a solution of $AlCl_3$ in $SbCl_3$ and uses HCl as a promoter. This process was first evaluated in an existing isomerization unit in 1961 (*26*) giving it the

Table VI. Summary of Published Data for Most Widely Used Isomerization Processes

Company[a]	*UOP*	*Shell Development*	*BP*
Catalyst	*Active Ingredients Supported on Spherical Base*	*Solution Of Al Cl_3 In Sb Cl_3*	*Chlorided Pt Al_2O_3 Pellets*
Operating Features			
Temperature, °F	250–400	150–210	320
Pressure, psig	200–1500	300	400
LHSV	1–4	2–3	2[f]
H_2 consumption, SCF/BBL	160	6[e]	
Promoter	None	HCl	HCl
Catalyst regeneration facilities	None	Yes[b]	Yes
Alloys for corrosion resistance	None	Yes	None
Conversions			
n-C_4→i-C_4, mole % i-C_4[c]	~60	62.5	N.A.
n-C_5→i-C_5, mole % i-C_5[d]	76–78	75.6	77
C_5/C_6 isomerization			
RON (clear) of charge	69.6[f]	—	60[f]
RON (clear) of product	81.9[f]	—	80.4[f]
Undesirable by-products formation			
C_4's and lighter, wt %	≤0.6	1.0[g]	≤0.5
C_7's and heavier, wt %	0	0.8[g]	0
Commercial installations	8	8	2
References	(*22,25*)	(*24,25,43*)	(*25,31,32*)

[a] Listed in order of commercial applications.
[b] In effect catalyst is regenerated through a scrubbing column to separate Al Cl_3 and Sb Cl_3 from a hydrocarbon sludge which is discarded.
[c] $100 \times i\text{-}C_4/(i\text{-}C_4 + n\text{-}C_4)$.
[d] $100 \times i\text{-}C_5/(i\text{-}C_5 + n\text{-}C_5)$.
[e] Not required for butane isomerization.
[f] Pilot unit data.
[g] For laboratory charge stock consisting of 30% (w) n-C_5, 45% (w) n-C_6, 25% (w) naphthenes.

advantage of being adaptable to some of the older but existing isomerization plants.

Independently and simultaneously, Texaco Inc. (*28*) and British Petroleum Co. (*29*) developed different versions of dual function catalysts having platinum on a chlorided alumina.

The commercialized BP and UOP processes employ different dual functional platinum catalysts. The BP process was first commercialized between 1963 and 1966; the UOP process in 1959. The BP catalyst has not been modified since its commercial introduction, but the UOP catalyst system has undergone further development (*24*). An advantage of the BP process is that the catalyst is activated and regenerated *in situ*. Considering the hydroscopic nature of all the isomerization catalyst systems, a regenerable catalyst offers the refiner greater ease during reactor loading and operating flexibility. It provides the means and ability to regenerate the catalyst instead of reloading it if a serious upset in the charge purification system occurs and prematurely deactivates the catalyst. Another difference is that the BP process requires HCl as an onstream promoter whereas the UOP process does not. Neither of these processes require any special alloys nor do they require special spent catalyst handling facilities as required in a Friedel-Crafts catalyst system.

The operating conditions for the three processes are very similar—only temperatures are somewhat dissimilar. The Shell Development system, employing a modified Friedel-Crafts system, operates at a lower temperature—150°–210°F *vs.* 250°–400°F for the other two processes. However, the equilibrium effects of the temperature differences are minimized as shown by the similarity in n-C_4 and n-C_5 yields shown in Table VI. Unleaded octane numbers for C_5/C_6 isomerate, obtained from a pure C_5/C_6 straight-run fraction, could not be found in the literature for the Shell process. However, pilot unit operations charging laboratory blends of n-C_5, n-C_6, and C_6 naphthenes have been reported (*26, 45*). In the Shell process the use of antimony trichloride and hydrogen has considerably reduced the amount of side reactions for a Friedel-Crafts system so that the yield for this process is quite close to the yield structure for the other two processes.

Isomerization Today. Present isomerization capacity in the Western Hemisphere and Western Europe is estimated to be about 50,000 BPD basis information provided in Refs. *26* and *34*. It is not known how many World War II plants may still be operating or are capable of being returned to service, but the number is believed to be very small. Near the end of World War II isomerization capacity peaked out at 45,000–50,000 BPD of which slightly over 40,000 BPD consisted of n-butane isomerization. All of these installations used one of five processes employing aluminum chloride complexes (*25*).

After the war the need for aviation alkylate declined rapidly, and most of the isomerization units closed down. During the motor gasoline octane race in the 1950's, a number of butane isomerization units were placed on stream. Several pentane isomerization units were placed on stream in the 1960's, and it is believed that only one or two plants today are being used to isomerize a C_5/C_6 straight run cut (*41*).

Future Trends for Alkylation and Isomerization

The strong growth trend for alkylation that has existed over the past 30 years in the United States and Canada is expected to continue as a result of the steadily increasing demand of gasoline. The gasoline market has grown about 4–5% per year over the last decade, and it is expected to continue to grow at about this same rate over the next decade (*41*). Alkylation has accounted for an increasing percentage of the gasoline production as shown in Figure 5. Therefore, if these trends continue, 40,000–60,000 BPD of new alkylation capacity will be added annually in the years ahead.

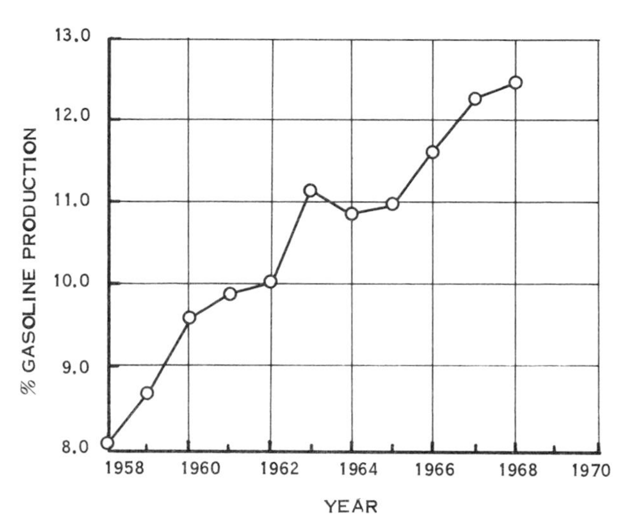

Figure 5. Total alkylation capacity as percentage of gasoline production

Gradual reduction or total elimination of lead anti-knock compounds would result in even higher annual growth rates for alkylation. How much higher will depend upon the unleaded octane level required to satisfy the lower compression ratios in future car engine designs. If the average unleaded RON for the gasoline pool stays below about 90 (*42*), then the predicted annual alkylation growth rate will remain at the estimated 40,000–60,000 BPD. In this case little additional isomerization capacity will be needed.

It now appears very likely that the average unleaded pool RON will exceed 90 within the next few years. Alkylation capacity will be increased proportionately, and the following trends for isomerization should emerge.

Butane Isomerization—C_4 isomerization will be limited to smaller refineries which do not contain hydrocracking facilities. Since the i-C_4/n-C_4 ratio for hydrocracking is about 2/1 to 3/1 sufficient isobutane should be available from this source to alkylate all of the available (C_3-, C_4-, and C_5-) olefins from catalytic cracking.

Pentane Isomerization—Isomerization of n-C_5 appears to be the most attractive for the C_4-C_7 paraffins. By recycling n-C_5 the unleaded RON for the light straight run pentane cut may be increased from 70-75 to 92.6. Assuming an average pentane fraction of 1.8 wt % (*43*) for crude, *n*-pentane isomerization has a potential capacity of about 200,000 BPD in the U. S. and Canada [assuming equilibrium conversion at 300°F, an average *i*-C_5/*n*-C_5 ratio in crude oil of 0.66 and an average refining capacity of 14 MM BPD]. This capacity could be increased somewhat by adding the pentane fractions from other units, such as the catalytic reformer and hydrocracker, to the isomerization unit deisopentanizer.

Hexane Isomerization—Isomerization of hexanes or a C_5/C_6 light straight run cut is predicted to be less popular commercially since the octane appreciation for the hexane portion is not as high as the 18-23 numbers predicted for *n*-C_5 isomerization alone. It is estimated that the comparable increase for hexane isomerization is about 10-15 octane numbers (RON, clear). This figure is more difficult to estimate as it is not only a function of hexanes and C_5/C_6 naphthenes in the crude source but it also depends on the degree of fractionation. Moreover, it is not unlikely that other means of upgrading the C_6 fraction, such as novel reforming or dehydrogenation methods, will be more attractive since the increase in octane number by these means should be considerably higher. However, the hexane fraction represents a larger portion of each crude than the pentane fraction does and depending upon individual cases, considerable interest may develop in C_5/C_6 isomerization in the U. S. and Canada as a result of removing lead anti-knock compounds from gasolines.

Heptane Isomerization—It is not expected that isomerization of a heptane fraction *per se* will be commercially feasible since straight run heptanes are a choice stock for catalytic reforming. The estimated equilibrium octane number (RON clear) for C_7 paraffins at 98°F is only about 82 using Rossini's equilibrium data. [These data have been checked experimentally at 98.2°F by G. M. Kramer and A. Schriesheim (*44*). Good agreement was obtained except for the 2,2-DMP and 3,3-DMP which underwent side reactions.]

In Western Europe it is expected that new isomerization capacity may exceed alkylation installations since naphtha availability generally exceeds demand. By selecting isomerization over alkylation the octane number of the gasoline pool may be increased without increasing the volume. Moreover, olefinic charge stock avails for alkylation are considerably smaller in Europe since there are fewer catalytic cracking units per refinery than in the United States and Canada. It is predicted that C_5, and to a lesser extent C_5/C_6 isomerization, will prevail over alkylation in Western Europe until more catalytic cracking units are installed and/or a shift in the demand for naphtha over fuel oil is experienced.

Up to this point only the elimination of lead anti-knock compounds has been considered. Legislation on the volatility, olefin content, or aromatic content of gasoline could also have a significant effect on the future of alkylation and isomerization. It is impossible to predict what legislation, if any, will be passed to regulate the properties of gasoline, but the directional effects of such legislation can be indicated:

Reduced Volatility—Volatility reductions would decrease gasoline volume and octane levels and produce excess light hydrocarbons. All gasoline producing facilities would, of course, have to be expanded to make up the volume loss, but greater increases in alkylation and isomerization capacity would be needed to raise octanes and utilize as much of the light hydrocarbons in gasoline as possible.

Olefin Reductions—Alkylation capacity would increase with any reduction in olefin content because it would provide a way to use the C_4 and some C_5 olefins in gasoline. New isomerization capacity would be needed if substantial quantities of C_5 and C_6 olefins had to be removed. The C_5/C_6 olefins would have to be hydrogenated and isomerized to make up part of the octane loss.

Reduced Aromatics—Substantial alkylation and isomerization capacity increases would have to accompany any reduction in aromatics content. With limits on aromatics concentration, alkylate and isomerate would be the major tools available for controlling octanes.

Thus, it is obvious that the growth of alkylation and isomerization could be accelerated by any legislation which requires drastic changes in the composition of gasoline.

Literature Cited

(1) Mrstik, A. V., Smith, K. A., Pinkerton, R. D., "Commercial Alkylation of Isobutane," ADVAN. CHEM. SER. (1951) **5**, 97.
(2) Birch, S. F., Dunstan, A. E., Fidler, F. A., Pim, F. B., Tait, T., "Saturated High Octane Fuels Without Hydrogenation, The Addition of Olefins to Isoparaffins in the Presence of Sulfuric Acid," *J .Petrol. Technol.* (1938) 303.
(3) "Alkylation and Isomerization," *Oil Gas J.* (Jan. 26, 1959) F-18.
(4) Payne, R. E., "Alkylation—What You Should Know About This Process," *Petrol. Refiner* (Sept. 1958) 316.
(5) "Impact of New Technology on the U.S. Petroleum Industry 1964–1965," p. 290, National Petroleum Council, 1967.
(6) "Alkylation Spree to Hit Peak in '64," *Oil Gas J.* (Sept. 8, 1958) 70.
(7) "Alkylate Scarcity Squeezes Refiners," *Ibid.*, (Aug. 4, 1958) 49.
(8) Stormant, D. H., "Refiners Off on Alkylation Spree," *Ibid.*, (June 27, 1965) 90.
(9) Stormant, D. H., "Fast Spurt Seen for Alkylation," *Ibid.*, (Feb. 8, 1965) 47.
(10) Cupit, C. R., Gwyn, J. E., Jernigan, E. C., "Catalytic Alkylation," *Petro/Chem. Engr.* (Dec. 1961) 47; (Jan. 1962) 49.
(11) Albright, L. F., "Alkylation: Chemical and Engineering Factors for Reactor Design," *Chem. Eng.* (July 4, 1966) 119.
(12) Albright, L. F., "Alkylation Processes Using Sulfuric Acid as Catalysts," *Ibid.*, (Aug. 15, 1966) 143.
(13) Albright, L. F., "Alkylation Processes Using Hydrogen Fluoride as Catalysts," *Ibid.*, (Sept. 12, 1966) 205.
(14) Li, K. W., Eckert, R. E., Albright, L. F., "Alkylation of Isobutane with Light Olefins Using Sulfuric Acid, Parts I and II," *Am. Chem. Soc., Div. Petrol. Chem., Preprints* (1969) **14** (3), B5.
(15) Albright, L. F., "Comparison of Alkylation Processes," *Chem. Eng.* (Oct. 10, 1966) 209.
(16) Nelson, W. L., "Petroleum Refinery Engineering," 3rd ed., p. 660, McGraw-Hill, New York, 1949.

(17) Goldsby, A. R., Beavon, D. K., "Alky Units Revamped for More Output," *Petrol. Refiner* (June 1959) 165.
(18) Goldsby, A. R., U.S. patents **3,234,301** (Feb. 8, 1966), **3,227,774** (Jan. 4, 1966), **3,227,775** (Jan. 4, 1966), **3,422,164** (Jan. 14, 1969), **3,428,705** (Feb. 18, 1969), **3,448,168** (June 3, 1969), and **3,462,512** (Aug. 19, 1969).
(19) "Additive Increases Yields in Alkylation," *Oil Gas J.* (July 26, 1965) 112.
(20) Nelson, W. L., "Great Improvements Shown in Alkylation," *Oil Gas J.* (Jan. 1, 1962) 101.
(21) Buiter, P., Van't Spijker, P., Van Zoonen, D., "Advances in Alkylation," *Proc. World Petrol. Congr., 7th,* (1967) **4,** 125.
(22) Grote, H. W., Western Petroleum Refiners Association Meeting, San Antonio, Texas (March 1958).
(23) Ridgway, J. A., Schoen, W., "Hexane Isomer Equilibrium," "Abstracts of Papers," 135th Meeting, ACS, April 1959, A-5.
(24) Erickson, R. A., Asselin, G. F., "Isomerization Means Better Yields," *Chemical Eng. Progr. No. 3* (1965) **61,** 53.
(25) Gunness, R. C., "Isomerization," ADVAN. CHEM. SER. (1951) **5,** 109.
(26) Ruedisulj, W. M. J., Evans, H. D., Fountain, E. B., "Shell's Liquid Phase Isomerization Process for C4/C6 Fractions," *Proc. World Petrol. Congr., 6th,* (1963), Sect. III, Paper 9, p. 1.
(27) Lawrance, P. A., Rawlings, A. A., "Advances in Isomerization," *Proc. World Petrol. Congr.* (1967), **4,** 136.
(28) U.S. patents **3,242,228** and **3,242,229** (March 22, 1966); British patent **976,941** (Dec. 2, 1964).
(29) British patents **953,187, 953,188,** and **953,189** (March 25, 1964); U.S. patent **3,218,267** (Nov. 16, 1965).
(30) U.S. Patent **2,999,074** (Sept. 5, 1961).
(31) U.S. patents **3,419,503** (Dec. 31, 1968) and **3,441,514** (April 29, 1969).
(32) Lannear, K. P., Arey, W. F., Perry, S. F., Schriesheim, A., Holcomb, H. A., "A New Isomerization Process," *Proc. World Petrol. Congr., 5th,* (1959) Sect. III, Paper 3, p. 29.
(33) O'May, T. C., "Low Temperature Isomerization of Hexanes," *Proc. World Petrol. Congr., 6th,* (1963), Sect. III, Paper 4, p. 17.
(34) *Hydrocarbon Processing* (1964) **43** (9), 171–180; (1966) **45** (8), 213–219; (1968) **47** (9), 171–174.
(35) Burbridge, B. W., Rolfe, J. R. K., "BP's Isom Process Goes Commercial," *Hydrocarbon Processing* (1966) **45** (8), p. 168.
(36) Giannetti, J. P., Sebulsky, R. T., "Chlorided Platinum-Alumina Low Temperature Isomerization Catalysts," *Am. Chem. Soc., Div. Petrol. Chem. Preprints* (1969) **14** (1) 97.
(37) U.S. patents **3,242,229** (March 22, 1966) and **3,440,301** (April 22, 1969).
(38) U.S. patents **3,285,990** (Nov. 15, 1966) and **3,318,820** (May 9, 1967).
(39) Canadian patents **762,980** and **762,981** (July 11, 1967).
(40) *Oil Gas J.* (1968) **66** (15), 54.
(41) Bachman, W. A., "Forecast of the Seventies," *Oil Gas J.* (1969) **67,** 160.
(42) Stormont, D. L., "Auto's Role as Major U. S. Air Polluter Near an End," *Oil Gas J.* (1970) **68** (8), 57.
(43) Smith, L. M., "Qualitative and Quantitative Aspects of Crude Oil Composition," *U.S. Bur. Mines, Bull.* **642** (1968) p. 47.
(44) Kramer, G. M., Schriesheim, A., "Heptane Isomerization," *J. Phys. Chem.* (1960) **64,** 849.
(45) Evans, H. D., Fountain, E. B., Ross, W. E., "Low-Temperature Isomerization of Pentane and Hexane," *Proc. Am. Petrol. Inst.* (1962) Sect. III, pp. 231–240.

RECEIVED May 14, 1970.

8

Recent Developments in Ethylene Chemistry

RALPH LANDAU

Halcon International, Inc., 2 Park Ave., New York, N. Y. 10016

G. S. SCHAFFEL

Scientific Design Co., Inc., 2 Park Ave., New York, N. Y. 10016

Ethylene has almost completely replaced acetylene for petrochemicals production. Production of acetaldehyde directly from ethylene is displacing oxidation of ethanol and hydration of acetylene. Direct hydration of ethylene is supplanting sulfuric acid hydration processes and fermentation for manufacture of industrial ethanol. Ethylene can be oligomerized to butenes, trimers, α-olefins, and higher molecular weight alcohols. Chlorination of ethylene to trichloroethylene and perchloroethylene is displacing the traditional use of acetylene. New oxychlorination processes to produce vinyl chloride intermediates are eliminating acetylene hydrochlorination. Direct vinylation of acetic acid is replacing production of vinyl acetate from acetylene. An important use for ethylene is to manufacture ethylbenzene; new alkylation techniques, coupled with new dehydrogenation methods to produce styrene, are increasing this use for ethylene.

Ethylene chemistry encompasses such a broad scope that this discussion must be limited to new developments which have industrial or commercial significance. Therefore, only recent developments related to the production of heavy organic chemicals (petrochemicals) from ethylene are considered.

Ethylene or Acetylene?

In reviewing recent trends in ethylene chemistry it is evident that ethylene continues to replace acetylene as an economically preferred fundamental raw material. As new ethylene plants are constructed, many having capacities around a billion lbs/yr, the cost of ethylene production has become more attractive. Furthermore, hydrocarbon cracking processes have been improved technically and economically. A recent trend toward the use of lower cost heavy petroleum fractions as feedstocks is also important. Although processes for producing acetylene from hydrocarbons also have been improved, there is growing evidence that where a choice exists for producing industrial chemicals from either ethylene or acetylene, ethylene will continue to dominate in the future.

Vinyl Acetate

This trend is exemplified by recent events in the commercial production of vinyl acetate. Historically, vinyl acetate was produced exclusively from acetylene by reaction with acetic acid in the vapor phase using a fixed bed catalyst. This acetylene-based process can be carried out at high yields in plants requiring a moderately low capital investment. Nevertheless, many research projects were initiated with the intent to use ethylene in place of acetylene. The first such process to reach industrial application was that developed by Imperial Chemical Industries in the United Kingdom, where a 30,000 ton plant was commissioned in 1966 using a liquid-phase catalytic reaction of ethylene with acetic acid (*1, 2*). Unfortunately, operation was hampered by many problems, most serious of which was corrosion. As a result, the process was abandoned in the United Kingdom in late 1969, and the plant is no longer operating (*3*). In the interim, however, ICI granted licenses to Celanese in the United States and to Showa Denko in Japan, and plants with certain design changes are being operated by these companies.

During the same period in Germany Farbenfabriken Bayer, in cooperation with Farbwerke Hoechst, developed a vapor-phase process, which avoids many of the possible corrosion problems. This process has not yet been practiced by Bayer on a large scale but has been licensed widely, especially in Japan (*4, 5*), where, as a result, acetylene is no longer used to produce vinyl acetate.

In the United States, Celanese has obtained a license from Bayer to expand their vinyl acetate production. In addition, National Distillers has developed an alternative vapor-phase process and will use it in a new plant nearing completion. The adoption by Celanese of the Bayer vapor-phase process has led to patent infringement litigation among National Distillers, Celanese, and Bayer (*6*).

Acetaldehyde

Catalysts used to convert ethylene to vinyl acetate are closely related to those used to produce acetaldehyde from ethylene. Acetaldehyde was first produced industrially by the hydration of acetylene, but novel catalytic systems developed cooperatively by Farbwerke Hoechst and Wacker-Chemie have been used successfully to oxidize ethylene to acetaldehyde, and this process is now well established (*7*). However, since the largest use for acetaldehyde is as an intermediate in the production of acetic acid, the recent announcement of new processes for producing acetic acid from methanol and carbon monoxide leads one to speculate as to whether ethylene will continue to be the preferred raw material for acetaldehyde (and acetic acid).

Chlorinated Solvents

Another important industrial chemical which has been produced historically from acetylene is trichloroethylene. This process involves the chlorination of acetylene to tetrachloroethane, which then is dehydrochlorinated to trichloroethylene. The present trend is toward the use of ethylene as a primary raw material. New developments, such as that carried out cooperatively by Detrex Chemical Industries and Scientific Design Co., have been instrumental in improving process economics for converting ethylene to chlorinated hydrocarbons such as trichloroethylene (*8*). This new process starts with the formation of 1,2-dichloroethane from ethylene, either by chlorination in the liquid phase or by oxychlorination using hydrogen chloride and molecular oxygen. Under proper operating conditions, the subsequent chlorination of dichloroethane can be controlled to produce predominantly tetrachloroethane. Catalytic dehydrochlorination of the latter yields trichloroethylene. Under other conditions the predominant chlorination product is pentachloroethane, which can be dehydrochlorinated to perchloroethylene, another valuable chlorinated solvent.

Another modification of the process can be used to meet the growing demand for 1,1,1-trichloroethane (methylchloroform). In this version, the chlorination of dichloroethane can be directed toward maximum production of 1,1,2-trichloroethane (*9*). This product when dehydrochlorinated yields vinylidene chloride, a widely used monomer. Hydrochlorination of vinylidene chloride yields 1,1,1-trichloroethane, a solvent of increasing importance.

It is also possible to use ethyl chloride, derived from ethylene and hydrogen chloride, as a starting material. Controlled chlorination of ethyl chloride produces 1,1,1-trichloroethane and 1,1,2-trichloroethane in an economically useful ratio. The 1,1,2-trichloroethane, when further chlorinated to tetrachloroethane and the latter is dehydrochlorinated, yields trichloroethylene.

Thus, starting with ethylene, a valuable series of chlorinated solvents can be produced by a process which exhibits more favorable economics than can be obtained by the use of acetylene, with the added advantage that process versatility results in a broader range of products.

Vinyl Chloride

Another major chlorinated hydrocarbon is vinyl chloride. For many years acetylene was the sole raw material for the production of vinyl chloride by a catalytic fixed bed vapor-phase process. This process is characterized by high yields and modest capital investment. Nevertheless, the high relative cost of acetylene provided an incentive to replace it in whole or in part by ethylene. The first step in this direction was the concurrent use of both raw materials. Ethylene was chlorinated to dichloroethane, and the hydrogen chloride derived from the subsequent dehydrochlorination reacted with acetylene to form additional vinyl chloride.

Further research resulted in processes for converting mixtures of ethylene, hydrogen chloride, and molecular oxygen to dichloroethane in high yields (*10*). This oxychlorination process has been adopted to such an extent that all large, modern vinyl chloride plants use only ethylene as a feedstock. Numerous oxychlorination processes have been developed, but those used most widely include processes developed by B. F. Goodrich, Stauffer Chemical, and Monsanto (together with Scientific Design). They differ primarily in type of catalytic reaction system employed—Goodrich using a fluidized bed, Stauffer using a fixed bed, and Monsanto using a modified fluidized bed.

Demonstration of the technical feasibility of producing mixtures of acetylene and ethylene by pyrolysis of hydrocarbons (Wulff process or Kureha process) has led to the manufacture of vinyl chloride from such mixtures. The acetylene component reacts selectively with hydrogen chloride to form vinyl chloride, the residual ethylene is converted to dichloroethane, and the latter is cracked to vinyl chloride, with the resulting hydrogen chloride being recycled. However, this type of process has not achieved the industrial importance of the "all-ethylene" type of process.

Ethylene Oxide

Ethylene not only competes with acetylene in the production of industrial organic chemicals but has many important uses of its own. For example, there have been recent developments in the production of ethylene oxide from ethylene. For many years, ethylene oxide was produced by the dehydrochlorination of ethylene chlorohydrin. With the discovery of the catalytic oxidation of ethylene to ethylene oxide and the widespread industrial application of this technology, no new large ethylene oxide plants use the chlorohydrin process.

In direct oxidation plants, air has been the preferred oxidant for a long time. A significant recent development has been the use of oxygen in place of air (*11*). It has long been recognized that the capital investment for ethylene oxide plants could be reduced by using oxygen rather than air, but this advantage was cancelled by the price of oxygen or the capital investment associated with oxygen-producing facilities. As technology for the design and operation of air-separation plants has improved and as these plants have been constructed at large capacities, the cost of oxygen has fallen. Therefore, now that oxygen is available at lower prices, either in pipeline supply from large producers or from integrated oxygen plants associated with high capacity ethylene oxide plants, it has become the preferred oxidant. In cases, however, when the ethylene oxide plant is too small to justify an independent oxygen plant or is in a location where pipeline oxygen is not readily available, the use of air continues to be economically attractive.

Styrene

Styrene, one of the world's major organic chemicals, is derived from ethylene *via* ethylbenzene. Several recent developments have occurred with respect to this use for ethylene. One is the production of styrene as a co-product of the propylene oxide process developed by Halcon International (*12*). In this process, benzene is alkylated with ethylene to ethylbenzene, and the latter is oxidized to ethylbenzene hydroperoxide. This hydroperoxide, in the presence of suitable catalysts, can convert a broad range of olefins to their corresponding oxirane compounds, of which propylene oxide presently has the greatest industrial importance. The ethylbenzene hydroperoxide is converted simultaneously to methylphenylcarbinol which, upon dehydration, yields styrene. Commercial application of this new development in the use of ethylene will be demonstrated in a plant in Spain in the near future.

Another interesting recent development in styrene technology which will affect future consumption of ethylene relates to new methods for increasing conversion in the dehydrogenation of ethylbenzene. Several years ago, Scientific Design pioneered a technique for increasing conversion in this reaction. The net result was a marked decrease in the capital investment required for styrene plants. The present trend is toward even higher conversions by using oxidative dehydrogenation. This process involves the introduction of compounds having the ability to oxidize hydrogen selectively so that reaction equilibrium is displaced. Although no industrial application of oxidative dehydrogenation has yet taken place, process development is in an advanced stage, and it is expected that this method will offer a significant improvement in this important use for ethylene.

Polyethylene

Ethylene has long been important as a monomer. The growth of its use in the production of polyethylene—both high density and low density —has been dramatic. An interesting recent development in this use for ethylene has occurred in the field of low density polyethylene. This product is manufactured commercially in two types of reaction systems—high pressure autoclaves or tubular reactors. Characteristics of polymers depend upon the system employed in their production, so both methods have participated in the rapid growth of polyethylene. Products made in tubular reactors have certain advantages in the production of high clarity films, but such reactors formerly had limited capacity because of heat-transfer problems in a highly exothermic polymerization. Since ethylene decomposes explosively at high pressures and temperatures, careful temperature control must be exercised. New high strength construction materials have been developed which allow the thickness of reactor walls to be reduced to improve heat transfer. More importantly, new and more sophisticated process techniques have evolved so that conversion, and therefore capacity, can be increased. A few years ago a capacity of 75–100 million lbs of polymer per year per tubular reactor was considered to be the ultimate. Recently it has been demonstrated that over 200 million lbs/yr can be produced in a single reactor, retaining excellent polymer properties. Since capital charges are an important component of polyethylene production cost, these capital investment reductions have been an important development in the industrial utilization of ethylene.

Ethylene Oligomers

In addition to ethylene polymers, oligomers of ethylene have attained importance in the recent past. Linear primary alcohols and α-olefins in the C_{12} to C_{18} range, produced by variations of the Ziegler "Aufbau reaction," are industrial chemicals used in large quantities to produce biodegradable detergents. Furthermore, lower molecular weight oligomers are potentially important chemical intermediates. One example is the result of a recent joint development by Scientific Design and the Central

Institute for Industrial Research of Oslo, Norway. A unique metallo-organic catalyst system has been discovered which enables the selective trimerization of ethylene to 3-methyl-2-pentene. High temperature demethanation of this compound results in the formation of isoprene in good yields. Similarly, since 1-butene and 2-butene are dimers of ethylene, they react with ethylene selectively in the liquid phase to produce 3-methyl-2-pentene.

Isoprene has been produced commercially in a plant owned by Goodyear, using a process originally developed jointly by Goodyear and Scientific Design, through which 2-methyl-2-pentene is produced from propylene and then demethanated to isoprene. Since the yield of isoprene by demethanation of 3-methyl-2-pentene is higher than that from 2-methyl-2-pentene, ethylene can serve as an economical source for isoprene production.

Ethanol

Traditionally, ethanol has been made from ethylene by sulfation followed by hydrolysis of the ethyl sulfate so produced. This type of process has the disadvantages of severe corrosion problems, the requirement for sulfuric acid reconcentration, and loss of yield caused by ethyl ether formation. Recently a successful direct catalytic hydration of ethylene has been accomplished on a commercial scale. This process, developed by Veba-Chemie in Germany, uses a fixed bed catalytic reaction system. Although direct hydration plants have been operated by Shell Chemical and Texas Eastman, Veba claims technical and economic superiority because of new catalyst developments. Because of its economic superiority, it is now replacing the sulfuric acid based process and has been licensed to British Petroleum in the United Kingdom, Publicker Industries in the United States, and others. By including ethanol dehydrogenation facilities, Veba claims that acetaldehyde can be produced indirectly from ethylene by this combined process at costs competitive with the catalytic oxidation of ethylene.

These are a few of the recent developments in ethylene chemistry which have, or will have, an impact on the production of tonnage organic chemicals. As larger ethylene capacity is installed, greater impetus is given to further replacement of acetylene as a raw material, and the search for new or improved uses for ethylene becomes intensified.

Literature Cited

(1) Reis, Thomas, "Compare Vinyl Acetate Processes," *Hydrocarbon Processing* (1966) **45** (11), 171.

(2) "Acetic Costs Determine Best Route to VA," *European Chem. News* (Aug. 23, 1968) 28, 29.
(3) "VA Initiative Lies With Acetic Acid Producers," *European Chem. News* (Sept. 26, 1969) 36.
(4) Remirez, Raul, "Ethylene or Acetylene Route to VA," *Chem. Eng.* (Aug. 12, 1968) 94, 95.
(5) "Hoechst-Bayer Route to VA Cuts Raw Material Costs," *European Chem. News* (April 14, 1967) 40, 42.
(6) "Bayer Enters New Wrangle With National Distillers," *European Chem. News* (Aug. 14, 1970) 6.
(7) "Acetaldehyde Cost Same as Ethylene Price," *European Chem. News* (Dec. 7, 1962) 29.
(8) "New Life For Old Times," *Chem. Week* (Aug. 9, 1969) 30, 31.
(9) Fox, S. N., Simon, R. W., "Detrex-SD Offer Ethylene Route to Chlorinated Solvents," *European Chem. News* (Sept. 15, 1970) 86, 88, 90.
(10) Burke, Donald P., Miller, Ryle, "Oxychlorination," *Chem. Week*, August 22, 1964, Pages 93–118.
(11) Landau, Ralph, Brown, David, Saffer, A., Porcelli, J. V., Jr., "Ethylene Oxide Economics—The Impact of New Technologies," *Chemical Eng. Progr.* (March 1968) 27.
(12) "Hydroperoxides Give Olefins," *Chem. Eng. News* (April 10, 1967).

RECEIVED May 25, 1970.

9

Ethylene Raw Materials and Production Economics

J. G. FREILING, C. C. KING, and J. NEWMAN

The Lummus Co., Bloomfield, N. J. 07003

The outlook for ethylene demand in the next decade is for continued strong growth with total new construction for the period forecasted at almost 100 billion pounds per year requiring the erection of about 100 plants. Careful attention will have to be given to proper selection of feedstocks if these huge investments are to be profitable. This paper discusses the differences relating to the various potential feed materials with respect to product distribution, investment and operating costs and examines the economics of ethylene production for a variety of situations. Both the U.S. and European pictures are covered, with emphasis on the effects of the important variables of feed and byproduct prices, operating severity, and scale. Future trends in feedstock availability and pricing are discussed as well as possible effects of a potential change in the U.S. import quota system and of unleaded gasoline.

The 1970s promise a continuation of the rapid worldwide growth in ethylene demand witnessed in the past decade. To meet the challenge of such growth successfully, it will be vital for planners to have a thorough understanding of the factors entering into the ethylene production economics.

Perhaps the most important of these factors involves the raw material employed for this purpose and the by-product volumes and prices. In this connection we discuss the product distributions from potential various feedstocks and current trends in feedstock selection, illustrating the significant role feedstocks play in the ethylene commercial picture. In addition, the effects on production economics of the factors of plant size and severity of operation are investigated.

Much of the information represents an updating of material presented in two papers related to the ethylene area prepared by the Lummus Co. in 1968 (*1, 2*). In addition, however, particular emphasis is given to possible new developments in the United States which could dramatically affect raw material choices. We explore the effects that a potential change in the import quota system and a move to lead free gasoline could have on the U.S. ethylene industry.

Ethylene Demands

The outlook for ethylene demand in the next decade is one of continued strong growth. Based on our discussions with several ethylene producing and consuming companies, we feel that the projections shown in Table I, which presents estimated annual ethylene demands for the 1970-80 period, are realistic and if anything, somewhat conservative.

In the United States ethylene demand is expected to go from 15.7 billion pounds in 1970 to over 35 billion pounds in 1980—a compounded annual growth rate of about 8½%. In Western Europe annual demand will grow by some 17 billion pounds between 1970 and 1980 from the 1970 level of about 12 billion pounds. An even more vigorous growth is predicted for the ethylene industry in Japan. Demand there will nearly triple in the next decade to almost 15 billion pounds by 1980 for a growth rate of nearly 11% per year.

Table I. Ethylene Demands—1970 to 1980

Billion Lbs/Yr

Year	*1970*	*1975*	*1980*
Location			
U.S.A.	15.7	24.7	35.5
W. Europe	12.0	19.5	29.0
Japan	5.3	9.6	14.8
Other (Ex. Eastern Bloc)	2.0	6.7	16.7
Total	35.0	60.5	96.0

The totals shown indicate that over-all annual demand (Ex. Eastern Bloc) for ethylene will grow by 62 billion pounds by 1980 corresponding to about a 10% per year growth.

To satisfy the increases in demand in the next decade, ethylene plant construction will have to continue at a brisk pace. Table II gives our estimates of plant growth and the inside battery limits (ISBL) capital investments required. An approximate 69 billion lbs/year increase in plant capacity must be installed in the period to meet the increase in demand. An additional 25 billion lbs will probably be installed to retire obsolete and uneconomic units. Thus, the total new construction for the decade will be *ca.* 94 billion lbs/year of ethylene. This will require the erection of about 100 plants. The associated order of magnitude ISBL battery limits investment for these plants will be around 4 billion dollars (current basis) for the surveyed areas.

Needless to say, the 1970s will require the careful selection of feedstocks by producers and the close cooperation of both producers and engineer-contractors to ensure that the challenge of such large investments can be profitably met.

Feedstock and By-Product Effects

The U.S. ethylene industry has been based primarily on the cracking of ethane and propane derived from natural gas. The quantities and liquid contents of U.S. natural gases have been such as to permit substantial quantities of these light hydrocarbons to be recovered for use as economically attractive ethylene feedstocks. In Europe and Japan, however, naphthas have been generally the available and preferred feeds to pyrolysis.

The light hydrocarbons produce only minor amounts of by-products, while naphtha and heavier feeds produce substantial quantities of propylene, butadiene, and aromatics. Thus, while in the United States these products are obtained generally from other routes at present, in Europe and Japan ethylene production serves as a major source of these chemicals. As discussed in greater detail later, by-product outlet considerations can play an important role in feedstock selection, and by-product realizations can have a major effect on the ethylene production economics.

Table II. Growth of Ethylene Industry—1970 to 1980

Location	*U.S.A.*	*W. Europe*	*Japan*	*Other (Ex. Eastern Bloc)*	*Total*
New demand, billions of lbs/yr	20	17	10	15	62
Equivalent capacity, billions of lbs/yr	22	19	11	17	69
Existing capacity to be retired, billions of lbs/yr	12	6	3	4	25
Total new capacity needed, billions of lbs/yr	34	25	14	21	94
Approx. no. of plants	28	25	16	30	99
Approx. total new ISBL plant investment, MM$	1200	1100	600	1000	3900

Production Patterns

Table III gives a range of the possible feedstocks that can be used to produce ethylene and the kinds and amounts of by-products that can be made from them. For our purposes we have selected a constant basis of 1 billion lbs/year ethylene production. The feedstocks illustrated in Table III include ethane, propane, *n*-butane, a full range naphtha, a light gas oil, and a heavy gas oil. The yields reflect high severity conditions with recycle cracking of ethane in all cases. For propane feed, propane recycle cracking has been included as well.

The figures for the naptha case pertain to a straight run C_5/360°F (TBP) cut point material; the light gas oil included here is a 386/633°F (ASTM) end point atmospheric distillate, and the heavy gas oil is representative of a light vacuum distillate with TBP cut point extending to 850°F. The once-through ethylene yields for these three feeds are 27, 23, and 21 wt %, respectively. Note that ethane recycle increases the yield of ethylene over the once-through figure. This of course means that less feed can be used for the same ethylene production. With the naphtha feed, ethane recycle cracking increases the 27 wt % once-through yield on feed to 31.4 wt %, reducing feed requirements by 16%.

As one progresses from ethane through heavier, lower hydrogen content feedstocks the yield of ethylene decreases and causes feed requirements to increase markedly. For example, cracking heavy gas oil would require slightly over three times the amount of feed (per lb of ethylene) that is needed for ethane cracking. The extra feed required for heavier stocks is, of course, distributed among the various by-products.

The total amount of by-products increases as feed molecular weight is increased. The heavier, lower hydrogen content feedstocks are more suited to *over-all* olefins production. Note especially the increase in the ratio of propylene/ethylene as the feedstocks become heavier. With ethane feed very little propylene is made, about 0.04 lb/lb ethylene whereas for the high severity propane cracking shown, the ratio is 0.39. For the heavier feeds the ratio tends to increase and reaches 0.57 for high severity cracking of the heavy gas oil. Interestingly, *n*-butane makes more propylene by-product than does this high severity naphtha operation and almost the same amount of propylene as does the heavy gas oil. Of course, operating at a lower severity will increase propylene/ethylene ratio.

In view of the predicted tight worldwide propylene supply/demand situation (*3*) the yield of propylene as well as that of ethylene will be an important consideration in feedstock selection.

As for the other by-products produced *via* cracking heavier feeds, the general trend is toward reduced off-gas production and increased pro-

Table III. Feed and Product

1000 MM Lbs/Yr

Feedstock	*Ethane*	*Propane*
Feed rate, 10^6 lbs/yr	1311.8	2379.6
Products, 10^6 lbs/yr		
Off Gas	211.7	713.7
Ethylene (polymer grade)	1000.0	1000.0
Propylene (chemical grade)	38.4	385.0
C_4 fraction		
butadiene	17.6	76.1
butylenes/butanes	7.5	32.6
C_5/400°F naphtha	36.6	142.9
400°F + fuel oil	—	29.3
Total products	1311.8	2379.6
BTX aromatics in pyro. Naphtha 10^6 lbs/yr	24.9	80.9
Hydrogen in feed, wt %	20.0	18.2

duction of butadiene, pyrolysis naphtha, and fuel oil. There is a marked increase in the 400°F+ fuel oil product when cracking gas oils. This will offer an important incentive for the ethylene producer to develop additional market outlets or uses for gas oil pyrolysis products. One route, providing for burning increased quantities of the material in the pyrolysis heaters, would sharply reduce, if not eliminate, the amount of fuel normally imported to gas oil plants. Other possible outlets for this material would be as carbon black or coker feed.

As discussed later, the aromatics produced by pyrolysis may take on added significance in the United States if the much talked about plans to remove lead from U.S. gasoline materialize. It is generally agreed that the values of aromatics as "process" octane improvers will be enhanced. Heavier feeds produce more aromatics-rich pyrolysis naphtha, with the full range naphtha yielding significantly more aromatics than any other feed considered in Table III. The concentration of BTX in the C_5/400°F fraction from naphtha cracking (at 27 wt % once-through ethylene) is about 65 wt %. This puts the yield of these aromatics at nearly 16 wt % of feed naphtha. Of course, to recover these aromatics the pyrolysis naphtha is usually hydro-treated first and then sent to aromatics extraction. The nonaromatic pyrolysis raffinate is highly naphthenic and is an excellent feedstock to catalytic reforming for production of additional aromatics.

Summary for Various Feedstocks

Ethylene Production

n-*Butane*	*Full-Range Naphtha*	*Light Gas Oil*	*Heavy Gas Oil*
2764.6	3193.6	3789.1	4108.1
592.4	578.7	453.0	482.3
1000.0	1000.0	1000.0	1000.0
564.3	470.5	537.5	568.1
84.1	119.5	147.0	150.9
348.5	133.2	177.0	177.8
112.0	771.9	739.6	727.0
63.3	119.8	735.0	1002.0
2764.6	3193.6	3789.1	4108.1
51.3	501.7	320.5	276.0
17.3	15.0	13.3	13.0

Alternative Feedstock Economics

Investments. The approximate inside battery limits (ISBL) investments for billion lb/year ethylene plants using the various feedstocks are given in Table IV. The figures presented represent a project getting underway on the U.S. Gulf Coast in early 1970. The investments of course could vary somewhat with the specific location and time of construction. They are based on production of propylene of only chemical grade and do not include pyrolysis gasoline hydro-treating nor butadiene or aromatics extraction facilities. The costs indicate a significant increase as the feeds become heavier.

Table IV. Approximate ISBL Plant Investments

1000 MM Lbs/Yr Ethylene Production

Feedstock	*Ethane*	*Propane*	n-*Butane*	*Full-Range Naphtha*	*Light Gas Oil*	*Heavy Gas Oil*
Approximate ISBL cost, MM$	35	40	42	43	50	52

Feed and By-Product Pricing Basis. Table V shows the feed and by-product price basis used in our economic calculations. Separate price structures are shown for the United States and Europe. For each case the by-product prices are assigned two values: (a) fuel value and (b) premium or chemical value. Pricing the by-products on each basis yields two values of ethylene production costs for each case considered. These

Table V. Assumed Feed

Feedstock	*Ethane*	*Propane*
Location		
Unit cost, ¢/lb	1.0	1.0
By-products		
Unit Values, ¢/lb	*Fuel Value*	
Off-gas	0.48–0.74	
Propylene (chemical grade)	0.41	
C_4 fraction		
butadiene	0.40	
butylenes/butanes	0.41	
C_5/400°F naphtha	0.38	
400°F + fuel oil	0.35	

values represent the maximum range of production costs one might expect under extremes of by-product valuation. A producer might be in an "overlap" situation—*i.e.*, some by-products can be sold at premium prices whereas others might only command fuel value. In that case, the ethylene production cost would fall somewhere between the two extremes mentioned above.

In setting the prices for the premium cases, we have incorporated a "sliding" price scale on some by-products; hence, a range of prices appears in some rows of Table V. These price variations reflect differences in the composition of a particular by-product which result from cracking different feedstocks. By way of example, the aromatics content of a pyrolysis naphtha depends on the specific feedstock from which it is derived. The premium price of the particular pyrolysis naphtha thus depends on its BTX concentration. The nonaromatic content of the pyrolysis naphtha is valued the same as naphtha. Further details can be found in Table VI.

The prices and values shown must be considered as illustrative only rather than as an attempt to predict the future. They are based generally on literature data averages and on discussions with a number of chemical and refining companies. While the figures used are generally representative of present price levels, in some cases a simplification is used—*e.g.*, taking ethane, propane, and butane prices at 1 cent/lb.

Effect of Feed and Value of By-Products on Production Costs. The ethylene production costs for the six feedstocks considered and the relevant by-product dispositions are shown in Tables VII and VIII.

and By-Product Values

n-Butane	*Full Range Naphtha*	*Heavy Gas Oil*	*Full Range Naphtha*	*Light Gas Oil*	*Heavy Gas Oil*
U.S.A.			Europe		
1.0	1.6	1.1	1.0	1.1	0.75

U.S.A.	Europe	
Premium or Chemical Value	*Fuel Value*	*Premium or Chemical Value*
0.73–2.84	1.05–1.06	1.40–1.45
2.75	0.89	2.0
5.0	0.86	4.0
1.30–2.20	0.87	1.0
1.30–2.10	0.81	1.05–1.30
0.50	0.59	0.59

Table VI. Further Details on Product Unit Values

	U.S.A.	*Europe*
Off-Gas		
A. Taken at fuel value, ¢/MM BTU(LHV)	21	45
B. Taken at prem. value		
contained H_2, ¢/MSCF	30	30
methane & other, ¢/MM BTU(LHV)	21	45
Butylenes/Butanes		
A. Taken at fuel value, ¢/lb	0.41	0.87
B. Taken at prem. value		
butylenes, ¢/lb	2.3	1.0
butanes, ¢/lb	1.0	0.87
Butadiene	—unextracted state—	
C_5/400°F naphtha		
A. Taken at fuel value, ¢/lb	0.38	0.81
B. Taken at prem. value untreated and before separation, based following values as finished products:		
benzene, ¢/gal	25	85% of U.S. values
toluene, ¢/gal	17	85% of U.S. values
C_8 aromatics, ¢/gal	19	85% of U.S. values
nonaromatics	—valued as naphtha—	

Table VII presents the data for ethane, propane, *n*-butane, full range naphtha, and heavy gas oil cases for the United States, while Table VIII shows the naphtha, light gas oil, and heavy gas oil based on a European situation.

The figures are based on:

(a) 1 billion lbs/year ethylene production.

(b) OSBL (outside battery limits) plant investment taken at approximately 25% of ISBL plant investment.

(c) Operating costs include direct labor, supervision, plant overheads, maintenance, utilities, catalysts and chemicals, local taxes, insurance, and depreciation.

(d) Return on investment taken at 20% before income taxes on total plant investment.

(e) Sales and startup expenses, working capital, spare parts, and general corporate administrative overheads are *not* included.

Table VII. Effect of Feedstock

1000 MM Lbs/Yr

Location	*U.S.A.*			
Feedstock	*Ethane*		*Propane*	
Costs, ¢/Lb Ethylene	*Premium By-products*	*Fuel By-products*	*Premium By-products*	*Fuel By-products*
Feed cost	1.31	1.31	2.38	2.38
Operating cost	1.16	1.16	1.32	1.32
By-product credits	(0.88)	(0.20)	(2.34)	(0.61)
Production cost (ex. return on investment)	1.59	2.27	1.36	3.09
Return on investment	0.88	0.88	1.00	1.00
Total production cost ¢/lb ethylene	2.47	3.15	2.36	4.09

Table VIII. Effect of Feedstock

1000 MM Lbs/Yr

Location		
Feedstock	*Full Range Naphtha*	
Costs, ¢/Lb. Ethylene	*Premium By-products*	*Fuel By-products*
Feed cost	3.19	3.19
Operating cost	1.68	1.68
By-products credits	(3.45)	(1.95)
Production cost (ex. return on investment)	1.42	2.92
Return on investment	1.08	1.08
Total production cost ¢/lb. Ethylene	2.50	4.00

Further details on the operating cost basis assumed are given in Table IX.

A number of conclusions can be drawn from Tables VII and VIII concerning feedstock choice. These are discussed below.

In the United States. With premium by-product prices prevailing, ethylene can be made from *n*-butane more cheaply than from either ethane or propane, assuming these light hydrocarbon feeds would be all available at 1¢/lb.

With fuel valued by-products in the United States, ethane is the preferred feedstock.

on Ethylene Production Costs

Ethylene Production

U.S.A.

n-*Butane*		*Full Range Naphtha*		*Heavy Gas Oil*	
Premium By-products	*Fuel By-products*	*Premium By-products*	*Fuel By-products*	*Premium By-products*	*Fuel By-products*
2.77 1.29 (3.07)	2.77 1.29 (0.76)	5.11 1.35 (4.13)	5.11 1.35 (0.92)	4.52 1.76 (4.53)	4.52 1.76 (1.23)
0.99	3.30	2.33	5.54	1.75	5.05
1.05	1.05	1.08	1.08	1.30	1.30
2.04	4.35	3.41	6.62	3.05	6.35

on Ethylene Production Costs

Ethylene Production

Europe

Light Gas Oil		*Heavy Gas Oil*	
Premium By-products	*Fuel By-products*	*Premium By-products*	*Fuel By-products*
4.14 2.16 (3.75)	4.14 2.16 (2.27)	3.08 2.28 (3.95)	3.08 2.28 (2.48)
2.55	4.03	1.41	2.88
1.25	1.25	1.30	1.30
3.80	5.28	2.71	4.18

Table IX. Basis for Operating Costs

	U.S.A.	*Europe*
Utilities		
Fuel, ¢/MM BTU(LHV)	21	45
Power, ¢/KWH	0.7	1.1
Steam, (H.P.), ¢/M lbs	65	110
Cooling water, ¢/M gal	2.5	3.0
Direct labor, $/yr./shift position	44,000	25,000
Supervision & plant overheads	—115%/yr of direct labor—	
Maintenance, material, & labor	4½%/yr ISBL + 2½%/yr OSBL	
Insurance & local taxes	2%/yr (ISBL + OSBL)	
Depreciation	10%/yr ISBL + 5%/yr OSBL	
Return on investment	10%/yr after 50% corporate taxes	

At current price levels, heavier feeds in the United States are not competitive with light hydrocarbon feeds. With U.S. naphtha at 1.6¢/lb (10¢/gal), the ethylene production costs from this feed ranges about 40–70% higher than costs associated with lighter feeds, assuming premium by-product values. The differences are even greater with fuel by-product values prevailing.

The current unattractiveness of heavier feeds in the United States notwithstanding, ethylene can be made more cheaply from 1.1¢/lb heavy gas oil than from 1.6¢/lb naphtha even though a gas oil plant is more expensive and requires more feed. This applies for both premium and fuel by-product cases. Even so, ethylene production costs with gas oil feed are about 25–50% higher than costs with light hydrocarbon feeds.

In Europe. With either premium or fuel by-product prices prevailing, naphtha is very marginally the preferred feedstock. Heavy gas oil appears to be an interesting feed possibility at current price levels. Although the ethylene production cost with this feed is slightly higher than with naphtha, the difference is so small that it could be wiped out by a naphtha price increase of less than 0.1¢/lb.

The conclusions advanced here depend on the feed prices as listed in Table V. The effect of feed price variations is very important in ascertaining the most profitable route to ethylene. We discuss this question later.

Another point to emphasize is that the economics presented in Tables VII and VIII are "ideal," *i.e.*, they do not reflect start-up and sales expenses, possible production delays, and other factors which would increase the cost of making ethylene above the figures shown.

Effect of Scale

Plant size is important in ethylene production economics as in most manufacturing operations. The economy of scale for ethylene stems basically from the reduction in capital requirements per unit of ethylene produced. This reduction in unit investment is a direct result of the well-known fact that as capacity increases, total plant investment does not go up in the ratio of capacities but rather in the ratio of capacities to some exponent which generally is less than one. Figure 1 shows the effect of variations in plant size on the investment per unit ethylene capacity.

Table X shows the effect of capacity on ethylene production costs for a European naphtha plant. Note that investment-based items such as return on investment and depreciation and others included in operating cost decrease, per unit of ethylene produced, as plant size gets larger.

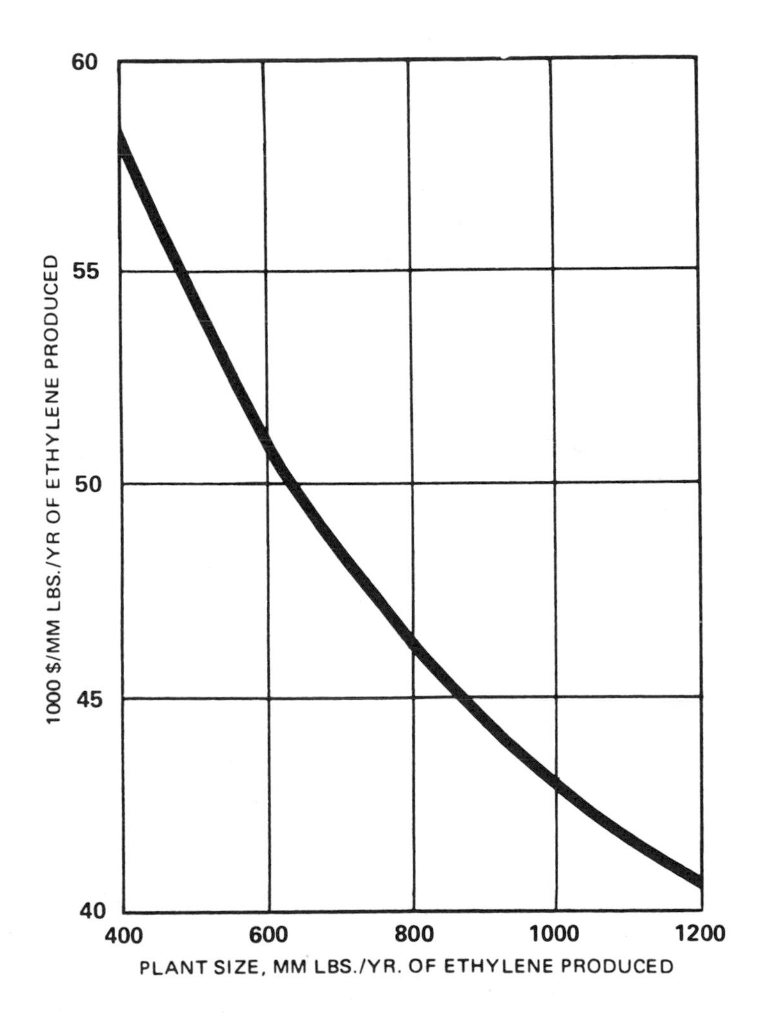

Figure 1. Effect of capacity on on-site investment (naphtha feed)

The ethylene production cost for a 1000 MM lb/yr plant is 2.5¢/lb (this figure also appears in the European naphtha column of Table VIII). For a 20% increase in capacity to 1200 MM lb/yr, the ethylene production cost drops to about 2.4¢/lb. For a decrease in capacity of 60% from 1000 to 400 MM lb/yr, the production cost increases by almost 30% to 3.2¢/lb. Table X shows that the advantages of scale diminish drastically as capacity is increased. By way of example, the decrease in production cost going from 400 to 700 MM lb/yr is about 0.5¢/lb C_2; the next 300 MM lb/yr increment (to a 1000 MM lb/yr) brings only a 0.2¢/lb. reduction in production costs.

Table X. Effect of Plant Capacity on Production Costs

Basis: Naphtha Feedstock at European Location;
Premium By-Product Values

Capacity, MM lbs/yr ethylene	*400*	*700*	*1000*	*1200*
Costs, ¢/lb. of ethylene				
Feed cost	3.19	3.19	3.19	3.19
Operating cost	2.03	1.80	1.68	1.62
Subtotal	5.22	4.99	4.87	4.81
By-product credits	(3.45)	(3.45)	(3.45)	(3.45)
Production cost	1.77	1.54	1.42	1.36
Return on investment	1.45	1.20	1.08	1.01
Total production cost	3.22	2.74	2.50	2.37

The other point to keep in mind is that as plant sizes become larger, the cost of the feedstock becomes a larger percentage of the total production costs.

It is evident that with the trend to larger plants, the small producer is at a distinct disadvantage in that his costs, per unit ethylene, are higher than for the bigger producers. To sell his product competitively he may be forced to accept a lower than desired profit, or even go out of business. Of course, he can retire his small unit and get back into things with a new big unit.

The trend has been to larger plants. In recent years, plants worldwide have been averaging about 650 MM lb/yr, and the outlook is for them to get even larger. We may see a billion lb/yr average size plant in the near future, especially in the United States. ICI recently completed a year of continuous ethylene production following a very smooth initial start-up of their billion lb/yr plant in Wilton, England. Shell is in the process of bringing a 1.2 billion lb/yr cracker on stream at their Deer Park, Tex. location.

Plant sizes can get larger, but bigness has negative factors associated with it as well. A single line plant of 1.5 billion lbs/yr capacity is possible today, but beyond this it is doubtful that unit investment costs would be reduced significantly. Limitations in commercially proven compressor designs, the need to field fabricate the major fractionating towers, etc., would act to limit further important cost savings. The "super large plant" would also be more vulnerable to market uncertainties and "eggs in one basket" risk.

In this latter connection both contractors and owner management play a joint role in ensuring reliability of operation and avoidance of unscheduled shutdowns. Great strides have been made in this area with better design, inspection and safety procedures, and by the use of improved maintenance and operating techniques.

Effect of Severity

Let us look at changes in cracking severity and how they affect feed requirements, by-product production, and ethylene production costs. We have selected the once-through yield of ethylene as a convenient means of representing severity.

Yield Pattern. Table XI presents a feed/product summary for a naphtha based billion lb/yr ethylene plant at various severities of 23, 25, and 27 wt % ethylene (once-through basis). The naphtha feed is the same one as referred to earlier (see Table III). It is immediately apparent that feed requirements are increased at lower severities for a given ethylene production rate. Also, production of olefin by-products increases as severity decreases. Note especially the 36% increase in propylene production as severity is dropped from 27% ethylene to 23% ethylene. Butadiene production goes up somewhat, while butylenes production jumps by over 100% going from 27 to 23% ethylene.

Table XI. Feed and Product Summary for Various Severities

1000 MM Lbs/Yr Ethylene Production from Naphtha

Once-Through Ethylene, wt %	*23*	*25*	*27*
Naphtha feed rate, 10^6 lbs/yr	3728.6	3437.6	3193.6
Products, 10^6 lbs/yr			
Off-gas	524.5	552.0	578.7
Ethylene (polymer grade)	1000.0	1000.0	1000.0
Propylene (chemical grade)	639.0	569.5	470.5
C_4 fraction			
butadiene	153.0	138.0	119.5
butylenes/butanes	272.5	198.5	133.2
C_5/400°F naphtha	1055.0	880.0	771.9
400°F + fuel oil	84.6	99.6	119.8
Total products	3728.6	3437.6	3193.6
BTX aromatics in pyro. naphtha, 10^6 lbs/yr	515.0	500.0	501.7

Another glance at Table XI reveals that gas make is reduced when severity is dropped while BTX aromatics produced remains about the same over the severities studied. Since the amount of C_5/400°F produced increases at lower severities, the percent BTX in the C_5/400°F fraction decreases in order to retain a relatively constant aromatics production.

As cracking severity is increased, the tendency to form tars, polynuclear aromatics, coke, etc., is increased. A convenient measure of this tendency is the hydrogen content of the C_5+ produced. When this hydrogen content goes below 7%, cokes and tars are rapidly laid down in the heater and subsequent quench facilities thereby severely curtailing run lengths. Thus, maximum severity is set by this hydrogen limit. The 27% ethylene case of Table XI represents such a maximum for the naphtha in question. Of course, with feeds less rich in hydrogen—*e.g.*, gas oils—the 7% hydrogen limit is reached at ethylene yields lower than 27%.

Severity then, is an effective vehicle to vary the product distribution according to specific needs. This type of flexibility can be, and is often built into the design of an ethylene plant and adds relatively little to the cost of the plant. Another simple way to vary the by-product-to-ethylene ratio is to vary the extent of recycle cracking of ethane.

Production Costs. The effect of varying severity on ethylene production costs is given in Figure 2. Both premium and fuel by-product situations are covered. The curves apply to a billion lb/yr European naphtha pyrolysis plant.

With fuel by-product values, the cost of producing ethylene increases as once-through severity decreases from 27 wt %. For example, at 27 wt % ethylene, the production cost is 4.0¢/lb; at 23 wt % the cost rises to over 4.2¢/lb. This is primarily a result of more feed being downgraded to fuel at lower severity.

More interesting, however, is the result when premium by-product values apply. Here we see that the production cost dips through a minimum at about 25% severity. This may be explained by noting that as severity drops, by-product credits increase owing to increased olefins production so that by-products credits minus feed costs increase, even while feed requirements go up. This tends to lower production costs. On the other hand, the investment and operating costs tend to increase as severity drops—this tends to raise production costs. The net effect is a minimum in production costs at about 2.45¢/lb of ethylene.

Implicit in Figure 2 are changes in investments and utility costs associated with severity changes. For the naphtha plant at 1000 MM lb/yr capacity, the investment in going from 27% severity to 23% is increased by about 1.9 MM$; the utilities cost increases by about 0.5 MM$/yr.

Feed Availability and Effect of Feed Prices

Earlier we showed that variations in the prices for feedstocks could have an important influence in determining the best ethylene route. Now we quantify these effects and tie this into current trends regarding feedstock availabilities which would influence the question.

Europe. The demand for naphtha for petrochemical feed purposes in Europe continues to rise strongly. As a result, the prices for this commodity have risen over the past several years. Increased use of light North African crudes, providing a higher proportion of gasoline and potential petrochemical naphtha, has retarded but not stopped the rise in naphtha prices.

It would be foolhardy to attempt to predict definitively the level of naphtha prices at some particular future date. However, additional

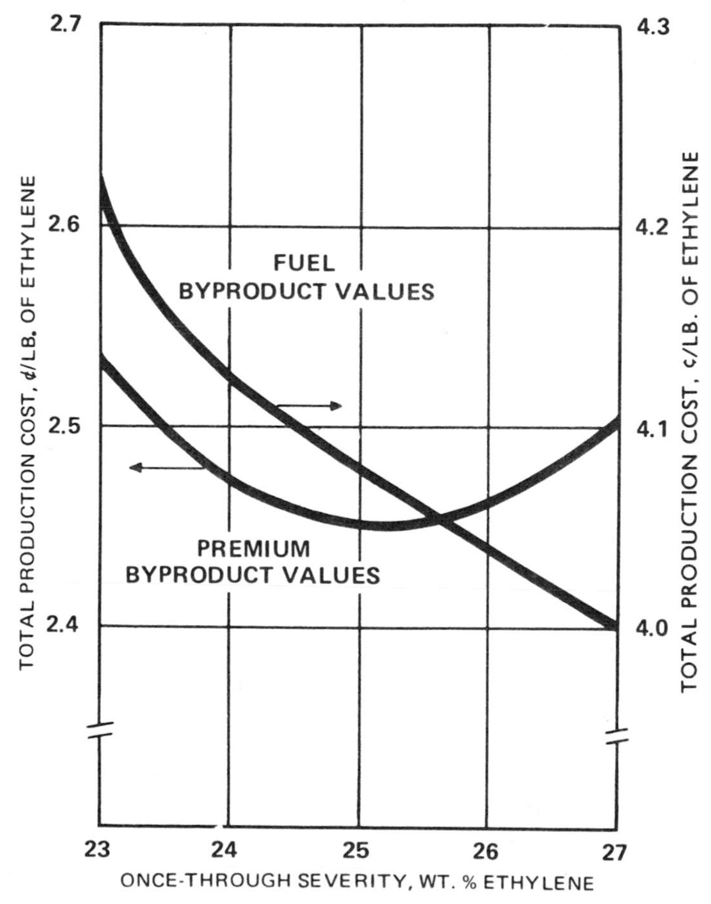

Figure 2. Effect of severity on European ethylene production costs (1000 MM lbs/yr ethylene production from naphtha feed)

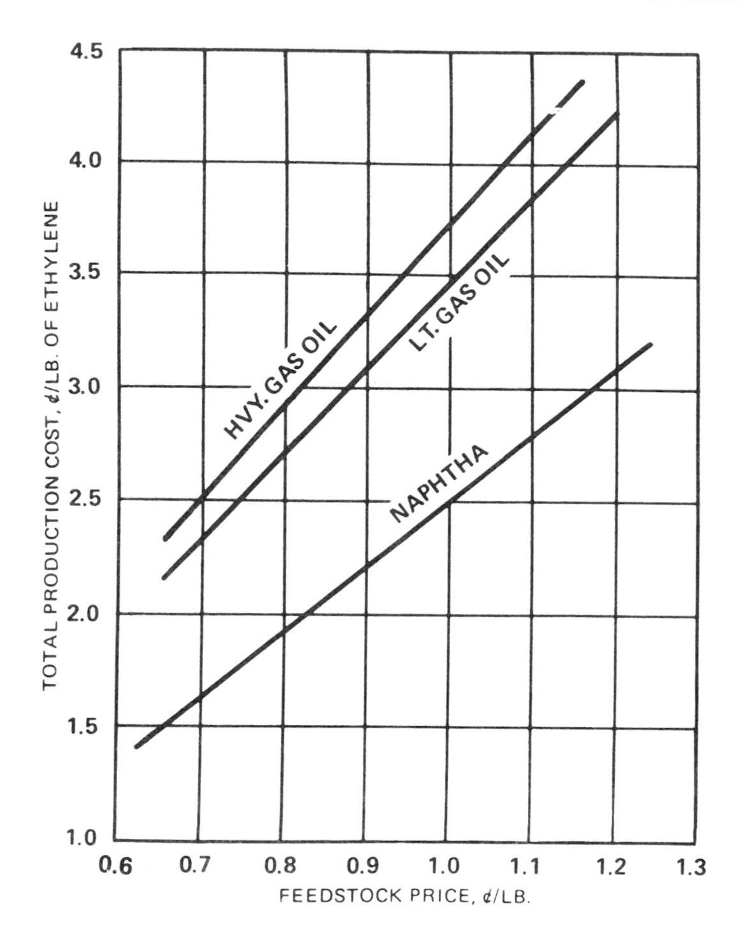

Figure 3. Effect of feedstock price on European ethylene production costs (1000 MM lbs/yr ethylene production; premium value by-products)

naphtha can be produced for petrochemical purposes in European refineries by practicing a more intensive degree of crude oil processing.

Installation of more conversion equipment both in new refinery construction and as additions to existing hydroskimming facilities is already a trend. The production of more naphtha by providing new conversion units would, of course, make the additional naphtha more costly. In this connection a number of studies both our own and others (*4*) have attempted to determine the cost of incremental naphtha production. These indicate that in typically sized European hydroskimming refinery (operating on either Libyan or Arabian crudes) gasoline plus petrochemical naphtha yields can be increased by about 50% by installation of catalytic cracking. Based on today's prices for the other refinery products, the cost of the incremental naphtha obtained in this manner would be about 1¢/lb (including a return on the investment required for the new facilities). Similarly an even larger increase in naphtha boiling range material would be feasible if hydrocracking rather than cat cracking were employed. Adding a hydrocracker could increase the base hydroskimmer's gasoline plus naphtha yield by almost 150%. The *average* cost for this much larger quantity of additional naphtha would be about 1.3¢/lb.

Figure 3 presents the effect of feed price on ethylene production costs in a billion lb/yr European plant for naphtha, light gas oil, and heavy gas oil feedstocks based on premium by-product valuations.

From this display it can be seen that a 0.1¢/lb increase in naphtha price from 1.0 to 1.1¢/ lb would raise the ethylene production cost from 2.5 to 2.8¢/lb. If such a movement did occur, heavy gas oil at current 0.75¢/lb levels (based on equivalent heavy fuel oil adjusted for viscosity considerations and vacuum distillation costs) would become competitive with naphtha.

This "breakeven" relationship between heavy gas oil (actually light vacuum gas oil) is shown more conveniently in Figure 4.

The breakeven price defines the unit price one could afford to pay for the gas oil to realize the same ethylene production cost as for a naphtha feed. If for a given naphtha price, the gas oil can be obtained at a price *below* the indicated gas oil breakeven price, gas oil would be the more attractive feed. If the gas oil can only be obtained at a price *above* the breakeven, naphtha would be the desired feed. Thus, with naphtha at, say, 1¢/lb, the heavy gas oil would have to be priced below 0.7¢/lb to be more profitable than naphtha as a cracking feed, assuming premium by-products prevail. Valuing the by-products as fuel has only a slight effect on the breakeven levels. The curves cross as nonaromatics in the pyrolysis gasoline have been valued the same as naphtha feed.

Thus, at today's naphtha prices and by-product markets, pyrolysis of light vacuum gas oil appears closely competitive with naphtha for ethylene production.

The breakeven curve for light (atomospheric) gas oil is given in Figure 5. If light gas oil can be disposed of at prevailing middle distillate prices of about 1.1¢/lb, naphtha would have to sell at 1.4 to 1.5¢/lb for the gas oil to become attractive. Therefore, pyrolysis of atmospheric gas oil will not be ordinarily attractive.

United States. The U.S. ethylene industry has been based mainly on pyrolysis of light hydrocarbons, predominately ethane and propane recovered from natural gas. Essentially all the ethane recovered is used as pyrolysis feed, whereas only about one quarter of the propane is used for this purpose. The remainder is mostly consumed in the LPG market.

Recent cold winters plus increased demands for petrochemical feedstocks have reversed an oversupply situation on the natural gas liquids to one where demand appears to be outpacing supply.

In view of the above, various sources predict price increases for natural gas liquids for petrochemical use (*5*, *6*). Let us look at the predicted magnitudes of these possible price increases and their effect on ethylene production costs.

One of the sources predicts that ethane prices will reach 1.1–1.3¢/lb and propane prices 1.2–1.4¢/lb on the Gulf Coast in the mid 1970's.

Figure 6 indicates the ethylene production costs from various feedstocks in a U.S. billion lbs/yr ethylene plant based on premium valued by-products. If the predicted ethane and propane increases did in fact

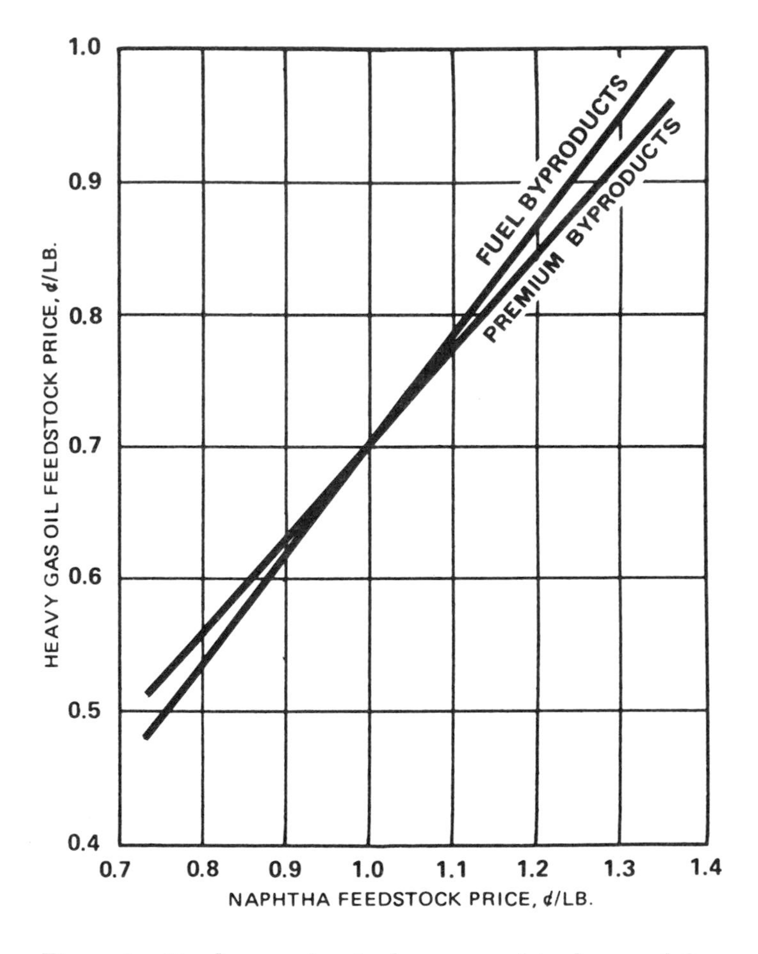

Figure 4. Breakeven prices for heavy gas oil feed vs. *naphtha feed in Europe (1000 MM lbs/yr ethylene production; premium and fuel value by-products)*

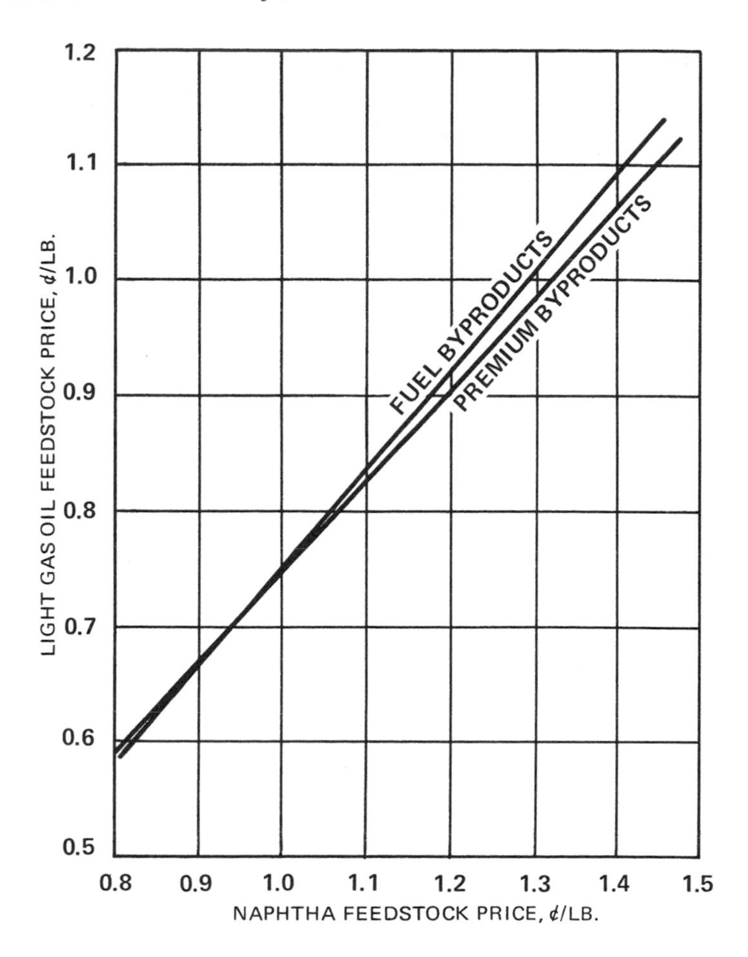

Figure 5. Breakeven prices for light gas oil feed vs. *naphtha feed in Europe (1000 MM lbs/yr ethylene production; premium and fuel value by-products)*

materialize, Figure 6 tells us that the cost of making ethylene from ethane will rise from 2.5¢/lb (with 1¢/lb ethane) to 2.6–2.85¢/lb and that cost from propane cracking will increase from 2.4¢/lb (with 1¢/lb propane) to 2.85–3.3¢/lb.

Predictions for *n*-butane prices are not readily available. The picture for *n*-butane is more complicated since large amounts of this material are also used as a blending stock in gasoline. Future trends in refinery uses of *n*-butane will certainly affect price levels. The current price for *n*-butane on the Gulf Coast is about 1¢/lb (*ca.* 5¢/gal). If the price goes to, say, 1.2¢/lb (*ca.* 6¢/gal), the ethylene production cost associated with *n*-butane cracking will rise to about 2.5¢/lb *vs.* 2.05¢/lb at the lower level.

With these types of increases in production costs arising from higher NGL prices, the heavier feedstocks would become much more competitive at the existing price valuations for U.S. naphtha, heavy gas oil, and by-products. Note from Figure 6 that if propane goes to 1.4¢/lb the ethylene production cost would go to 3.3¢/lb, whereas the cost from 1.1¢/lb heavy gas oil will be slightly less at 3–3.1¢/lb, and the cost from 1.6¢/lb naphtha would be about 3.4¢/lb.

Impact of Possible Developments

It is interesting to speculate on the effects that two developments which appear to be taking shape could have on the U.S. ethylene plant feedstock situation. These include the possibility of lower priced foreign feedstocks becoming more available to petrochemical producers and the ramifications associated with the move to lead free gasolines. Each of these would tend to increase the attractiveness of using heavier feedstocks, and the two acting together could have a significant impact on the future nature of the U.S. ethylene industry.

Imported Feeds. President Nixon has deferred action on the recommendations of the Cabinet Task Force on Import Controls pending consultations with foreign countries and the development of additional information in Congressional hearings (*7*). Whether or not the specific recommendations of the task force report regarding freer petrochemical producer access to foreign feedstocks are ultimately implemented, there seems to be a growing consensus on the likelihood of future changes of this nature coming to pass. Let us therefore examine the implications that this could have on the ethylene picture.

Much controversy has existed regarding the price at which foreign naphtha could be delivered in the U.S. and the question of whether foreign naphtha would be indeed available. Estimates of prices applicable on the Gulf Coast have ranged from about 5.3¢ to over 7.1¢/gal (0.85¢/lb to over 1.14¢/lb) (*8*). It is beyond the scope of this presentation to attempt to render judgment on the availability/price relationship that would result at a given world demand level. However, for purposes of illustration, let us look at what, say, a 1.1¢/lb naphtha price would mean in the United States.

Figure 6 shows that with the present level of premium valuation for by-products, a 1.1¢/lb naphtha price would result in this feedstock having an advantage over ethane, propane or butane at 1¢/lb. The cost for naphtha-based ethylene in this case would be only 1.94¢/lb *vs.* 2.04, 2.36, and 2.47¢/lb from *n*-butane, propane, and ethane, respectively. The breakeven prices for the light feedstocks that would correspond to the 1.1¢/lb naphtha price would be 0.6, 0.82, and 0.95¢/lb for ethane, propane, and *n*-butane, respectively. Thus the attractiveness of the ethane and propane feedstocks could be sharply impaired even at their current cost levels.

Although the imported cheaper naphtha feedstock has been used here for illustration, similar considerations would apply to enhancing the position of imported gas oil as well.

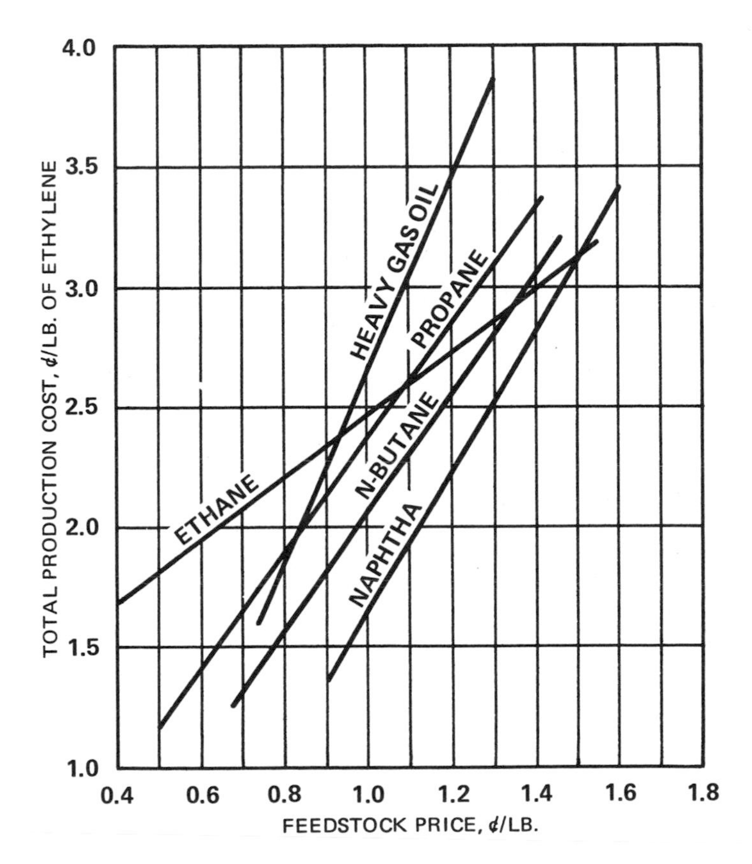

Figure 6. Effect of feedstock price on U.S. ethylene production costs (1000 MM lbs/yr ethylene production; premium value by-products). Note: by-product prices as given in Table V. However, for n-*butane feed, the butanes contained in the C_4 by-product are valued the same as* n-*butane feed.*

Lead Free Gasoline. The current rapidly developing move toward unleaded gasolines in the United States will have a further impact in enhancing the position of the heavier feedstocks.

At the time this paper was written many questions relating to removal of tetraethyl lead (TEL) from gasolines are still unsettled, but these relate principally to timing, octane levels, etc., and it seems clear that lead will be phased out. While the levels of the gasoline octane that will be generally required in the future has not yet completely crystallized, it appears highly probable that these will be higher than associated with today's blends on a "non-leaded basis." This will require additional refinery investments and higher processing costs.

The nature of the specific additional processing that will be installed will be specific to each individual refinery's base situation, but in general the effect will be to increase the value of several of the by-products associated with ethylene production. These include propylene, butylene, and the aromatics in the pyrolysis naphtha. The prices of the two olefins will tend to rise as additional alkylation units are installed to boost gasoline octanes, thus making these chemicals more valuable to the

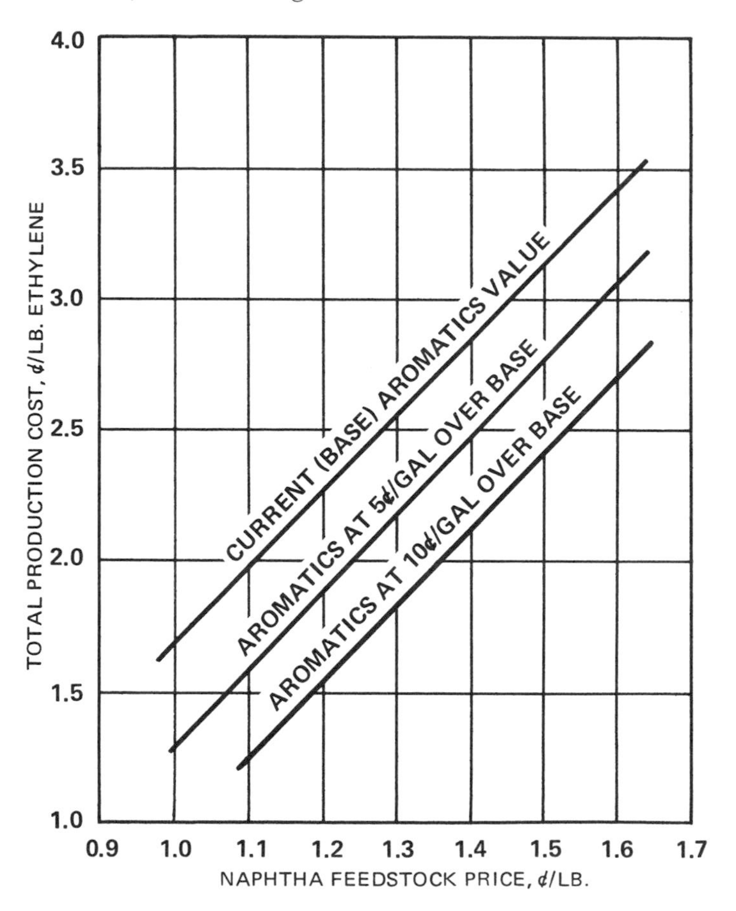

Figure 7. Effect of aromatics by-product price structure and naphtha feed price on U.S. ethylene production costs (1000 MM lbs/yr ethylene production from naphtha feed; premium value by-products)

Table XII. Basis for Determining Value of Aromatics-Rich Pyrolysis Naphtha

Location:	U.S.A.
Type Feed:	Naphtha
Severity:	27 wt % C_2^-, once-through

Value of aromatics-rich pyrolysis naphtha determined by value of its components as "finished" chemicals, and the costs incurred in treating and separation required to produce them from pyrolysis naphtha. Breakdown of pyrolysis naphtha is as follows:

	C_5/400°F Pyrolysis Naphtha Composition, wt %	*Finished Price as Chemicals (Base Value), ¢/gal*
Benzene	29.2	25
Toluene	21.0	17
C_8 aromatics	14.8	19
Nonaromatics	35.0	Same as naphtha feedstock (variable)
	100.0	

The cost incurred in treating and separating the pyrolysis naphtha into its constituent parts is approximately 0.6 ¢/lb of pyrolysis naphtha.

refiner. Aromatics prices will go up as increasing quantities are used in gasolines and are less available for petrochemical purposes. As discussed earlier, the heavier feeds, for a fixed ethylene production, would provide a considerably greater make of these higher value by-products than would the light feeds. Let's now look at the effect on the naphtha cracking economics of only one of these increased by-product values—the aromatics.

The study carried out by Bonner and Moore for the American Petroleum Institute in 1967 (*9*) on the costs of producing unleaded gasolines investigated the enhancement in value to the refiner of the BTX included in the gasolines. The value increases calculated by them as applying to a significant part of the total BTX product ranged from 9 to 15¢/gal. Their study was based on the elimination of TEL while maintaining current octane numbers. Since the current prospect is for the octane requirements to drop somewhat (*via* reduction of compression ratios), it is likely that only a fraction of 9 to 15¢/gal increase would apply. We will therefore consider aromatics price increases in the 5 to 10¢/gal range.

Figure 7 shows the effect on ethylene production cost from naphtha cracking with BTX aromatics value increases as a parameter. (The basis we have used for determining the effect of aromatics price increases is given in Table XII). Figure 7 indicates that a 5¢/gal increase in BTX value would reduce the ethylene production cost by about 0.35¢/lb (assuming other by-product valuations remained unchanged).

With imported naphtha at, say 1.1¢/lb and aromatics at current values, ethylene cost is 1.94¢/lb. However, with finished aromatics valued at 5¢/gal over the current base values, production cost drops to 1.59¢/lb. The breakeven curves for naphtha *vs.* ethane, propane, and *n*-butane are given in Figures 8, 9, and 10. These assume premium by-products, with aromatics valued above current levels, but do not include the effect of increased propylene and butylene valuations that would further accentuate the picture. With 1.1¢/lb naphtha and aromatics at 5¢/gal above current prices, the breakeven prices for ethane, propane, and *n*-butane are 0.33, 0.7, and 0.83¢/lb, respectively. Such prices are

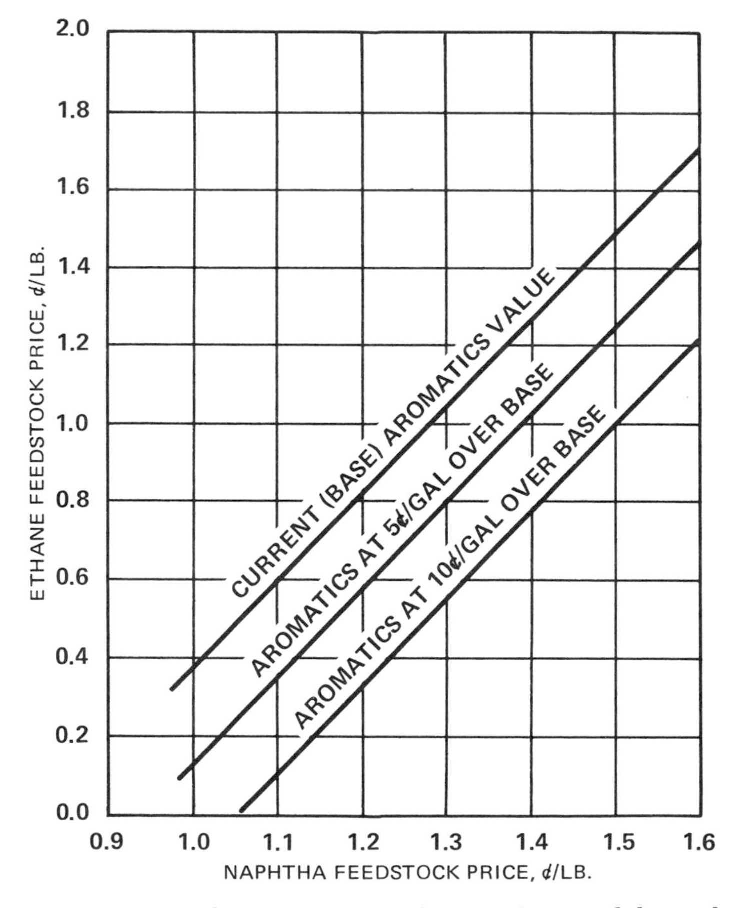

Figure 8. Breakeven prices for ethane feed vs. *naphtha feed in the United States. Parameter is aromatics by-product value. 1000 MM lbs/yr ethylene production; premium value by-products.*

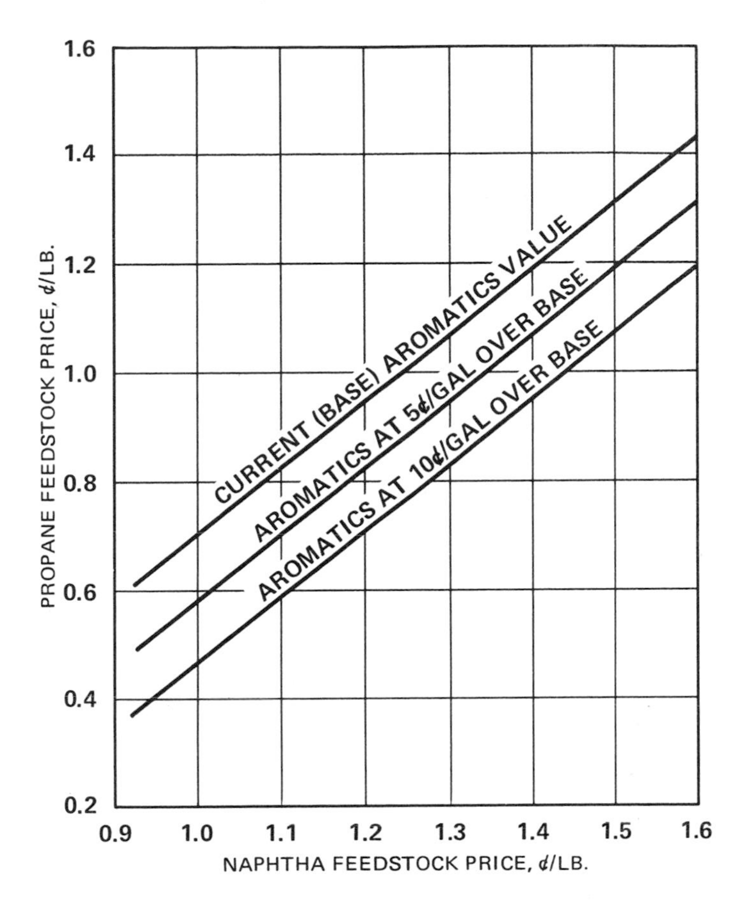

Figure 9. Breakeven prices for propane feed vs. *naphtha feed in the United States. Parameter is aromatics by-product value. 1000 MM lbs/yr ethylene production; premium value by-products.*

well below current levels for these hydrocarbons. In fact the ethane break even is even below its value as fuel of 0.43¢/lb.

Under the assumptions of naphtha price and aromatics value stated above, naphtha pyrolysis clearly would be superior to light hydrocarbon pyrolysis at their current feed prices. A similar analysis can probably be made also for gas oil. Thus, if the possible developments discussed above do materialize, the heavier feeds could probably dominate almost all new U.S. ethylene plant construction in the future.

Another interesting trend which may develop as a consequence of lead free gasoline is increased interest in isobutane pyrolysis. Hydrocracking may well be used as an important tool in the lead free gasoline refinery. This process produces considerable yields of isobutane. It is

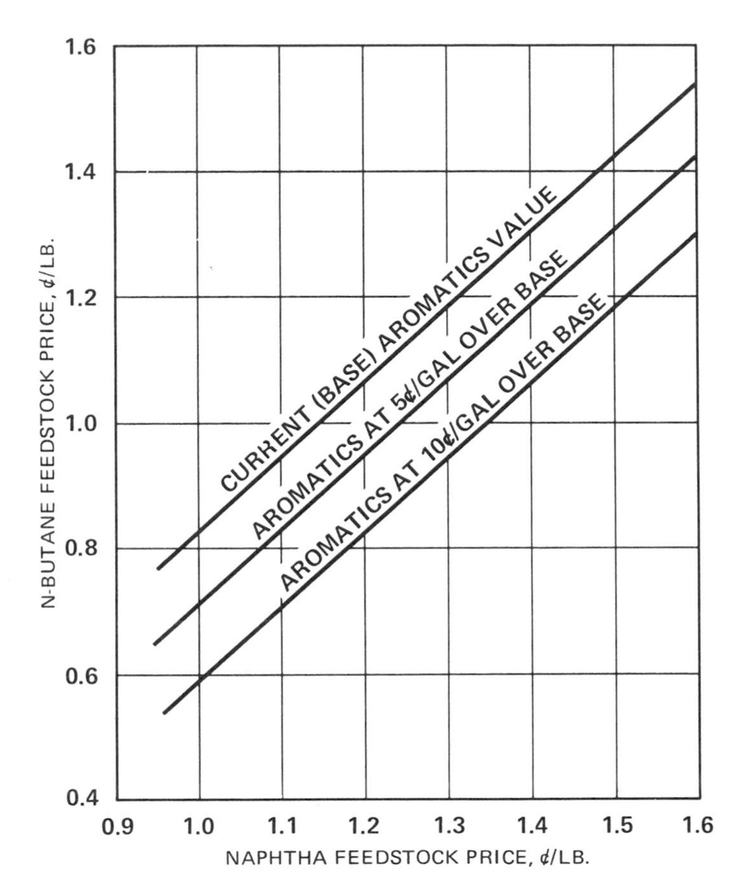

Figure 10. Breakeven prices for n-*butane feed* vs. *naphtha feed in the United States. Parameter is aromatics by-product value. 1000 MM lbs/yr ethylene production; premium value by-products.*

probable that if a refinery found itself with an excess of isobutane from hydrocracking (excess relative to propylene and butylene), isobutane pyrolysis could be employed to produce the additional olefins required for alkylation balance.

We stress again that our comments regarding a strong movement toward the heavier feedstocks—resulting from the possible developments related to lower cost imported stocks and lead free gasoline—must be regarded as speculative only at this time. This move to heavy feeds would however be the logical consequence of such developments maturing.

Conclusions

(1) The current decade's outlook is for continued strong worldwide growth in ethylene demand. Total new plant construction for the 1970's

will approximate almost 100 billion lbs of ethylene capacity requiring about 100 new plants.

(2) The many factors involved in the ethylene production economics are interrelated in a complex manner. The more important variables relate to feedstock types and prices, by-product volumes and valuations, plant size, and severity of operation.

(3) Currently in the United States, ethylene cost is lower *via* production from the light feedstocks regardless of the type of by-product valuation applicable. Of these, *n*-butane appears most interesting if premium prices for by-products can be applied. Ethane is best for limited by-product outlets.

(4) In Europe, cracking of either light vacuum gas oil or naphtha give very comparable ethylene production costs at current price levels. This holds for either premium or fuel value by-products.

(5) Plants are still getting bigger, but further important savings caused by larger scale would seem hard to come by.

(6) Operating severity can be used to vary product distribution to optimize the economics in the context of a producer's specific feed/by-product pricing situation.

(7) If widely predicted future price increases for the natural gas liquid feeds materialize, naphtha and gas on feedstocks will become more competitive in the United States even at current prices for the heavier feeds.

(8) Two possible future developments, should they become reality, could radically affect the U.S. ethylene industry. Lower cost imported heavier feedstocks and by-product value changes associated with lead free gasoline each would increase the attractiveness of using the heavy feeds. Acting together they would probably lead to future heavy feed domination of almost all new U.S. ethylene plant construction.

Literature Cited

(1) Freiling, J. G., Huson, B. L., Tucker, W., "Commercial Implications of Trends in Cracker Technology," European Petrochemical Association, 2nd Meeting, Knokke, Belgium, Oct. 1968.
(2) Freiling, J. G., Huson, B. L., Summerville, R. N., "Which Feedstock for Ethylene," *Hydrocarbon Processing* (1968) **47** (11), 145.
(3) Bland, W. F., "Petrochemical Growth Discussed in Wiesbaden," *World Petrol.* (1969) **40** (13), 34.
(4) Reis, T., "Europe Steers Its Own Course," *Oil Gas J.* (Dec. 22, 1969) 54.
(5) Aalund, L. R., "Ravenous Demand Spells Brighter Days for LP-Gas," *Oil Gas J.* (July 7, 1969) 77.
(6) "Chemco Seeks Unlimited Feed Imports," *Oil Gas J.* (June 30, 1969) 82.
(7) Kinney, G. T., "Nixon Shelves Tariff Plan, Asks Congressional Review," *Oil Gas J.* (March 2, 1970) 25.
(8) "Lifting of Naphtha Import Lid Opposed," *Oil Gas J.* (July 11, 1966) 66.
(9) Lawson, S. D., Moore, J. F., Rather, J. B., "Added Cost of Unleaded Gasoline," *Hydrocarbon Processing* (1967) **46** (6), 173.

RECEIVED July 30, 1970.

10

Vinyl Chloride Monomer and Poly(vinyl chloride)

L. A. DUWELIUS

B. F. Goodrich Chemical Co., 3135 Euclid Ave., Cleveland, Ohio 44115

H. R. SHEELY and S. T. SHIANG

The Badger Co., Inc., One Broadway, Cambridge, Mass. 02142

World-wide consumption of PVC [poly(vinyl chloride)] has increased dramatically in the past few years. It has now exceeded 8 billion lbs annually. The production of VCM (vinyl chloride monomer) has also been expanded to meet the PVC demand. Future trends for VCM and PVC productions for the next five years can be forecast on the basis of the raw materials sources, the different process techniques in manufacturing VCM and PVC, and their relative economics, technical merits, and limitations. VCM will be produced principally through the ethylene route by fluid-bed oxyhydrochlorination of ethylene and thermal cracking of ethylene dichloride. PVC will be produced by various processes resulting in more specialized PVC varieties tailored for specific end markets and new processing technologies.

Poly(vinyl chloride), like so many of the young basketball players of today, is "seven feet tall and still growing." From a fascinating laboratory discovery in 1926 it has grown to be the proved workhorse of the plastic industry. Consumption of PVC in the United States and the world has progressed steadily during the past two decades, increasing lately at the rate of 10% a year. As new applications continue to be developed and the price in comparison with other competitive materials continues to be attractive, it is expected that consumption will continue to increase at a high rate through the 1970's.

PVC Growth

Figure 1 shows how PVC consumption has grown in Europe (*1*), the United States (*2*), and Japan (*3, 4*) from the early 1950's to the present day and indicates that this trend is expected to continue. By 1975, each of these areas should be producing over 4 billion lbs/yr. Figure 2 shows how the average price of PVC in the United States (*2*) has dropped during the same period of time. These figures represent the average prices including specialty products in addition to the general purpose resin. The slight upward trend shown for the next five years is primarily a consequence of inflation.

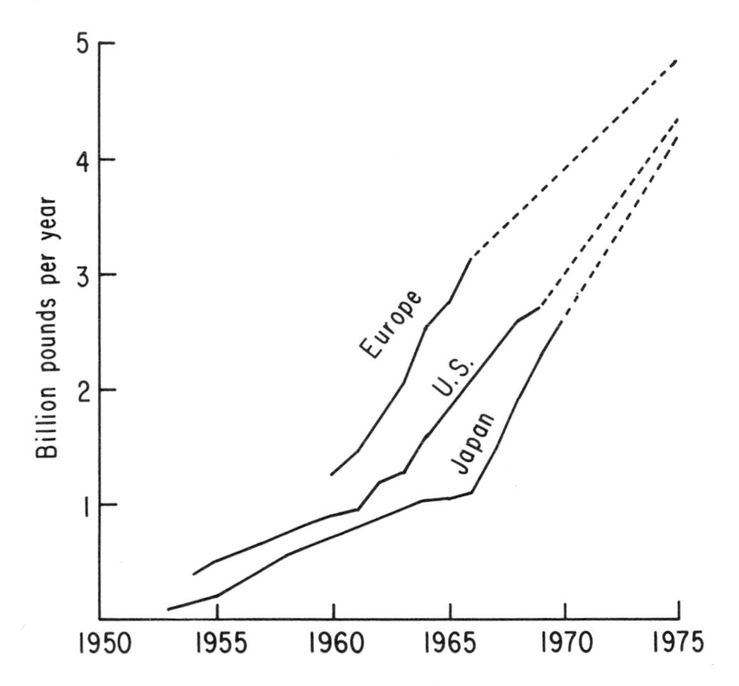

Figure 1. PVC consumption (based on data from Refs. 1, 2, 3, *and* 4)

To understand this dramatic increase in the use of PVC in competition with dozens of other plastic materials, we must look beyond the initial assumption that it has been solely a result of the declining price of the polymer. The spiraling consumption is primarily the result of the usefulness of the product. The declining price is the result of technological advances, mostly in the field of monomer production, plus increasing competition in polymer marketing. For example, in the United States the number of polymer producers has increased from six to 26 during the last 20 years. In other words, PVC is popular because of its versatility.

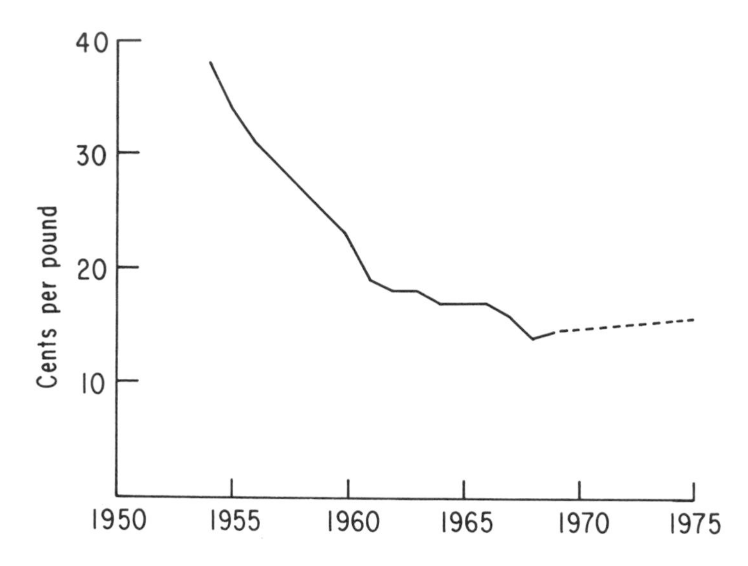

Figure 2. U.S. PVC price (based on data from Ref. 2)

Because of its popularity, manufacturers have been motivated to develop more efficient, and consequently cheaper, means of producing it.

PVC Applications

The extraordinary versatility of this polymeric plastic material, which has made it one of the most widely used of the myriad of polymers and copolymers available today, is demonstrated by its range of end uses. PVC products have properties that combine strength, lightness, and inertness with a wide range of resilience, excellent color acceptance, electrical insulating properties, crystal-like clarity, and the ability to be processed easily by many methods into a tremendous variety of forms.

Products made from PVC have properties ranging from the soft, resilient feel of a doll's cheek to the high impact strength of rigid piping. As a plastisol or organisol, PVC can be formed by rotational molding or slush molding, transformed to sponge or foam, or coated onto complex solid shapes. In a latex PVC is used in paints, sealants, and adhesives.

Figure 3 shows the end uses, arranged in categories, of the PVC produced in the United States (5). About 35% of the PVC resin is extruded into pipe, complex forms, or flat sheets. Another 32% is calendered into film and flooring materials. About 12% is injection-molded or blow-molded into packages, records, and other forms; a lesser amount is coated onto paper or fabric. Approximately 4% is exported. It is easy to see how the demand for this extraordinarily versatile material would stimulate great interest in its production.

VCM Production

Technological advances in the production of the vinyl chloride monomer (VCM) have contributed to the declining price of the polymer. Figure 4 illustrates this statement; the price of the vinyl chloride monomer (*1*) over a period of 20 years is plotted against two curves that represent the annual production of monomer made from two different bases, acetylene and ethylene. The classic acetylene route was the first to be exploited commercially, but its popularity has declined as more processes were developed that could utilize ethylene, a cheaper base.

The use of ethylene instead of acetylene as a base for the vinyl chloride monomer represents not only a savings in the cost of raw material but a reduction in manufacturing costs (*3, 6, 7*), as shown in Figure 5. In these graphs, the unit costs from three sources have been reduced to a common basis. Total manufacturing costs, as well as raw material costs, are lower with an ethylene base than an acetylene base. Operating costs—*i.e.*, manufacturing costs minus raw material costs—are higher with ethylene than acetylene.

The three routes shown in Figure 5 are the classic all-acetylene, an all-ethylene route, and a "balanced" route which utilizes a mixture of acetylene and ethylene. A review at this point of the three different routes to VCM will help to define the relationships existing between the cost factors.

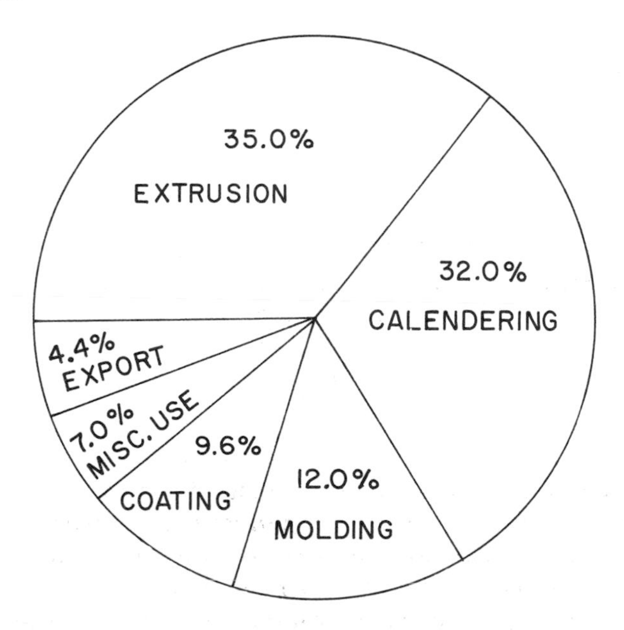

Figure 3. PVC end uses for the United States (based on data from Ref. 5)

All-Acetylene Route. The all-acetylene route, Figure 6, was the first to be developed. In 1937 Goodrich established a VCM and PVC pilot plant, and by 1940 it was producing PVC commercially at its Niagara Falls location. This process (*8*) was characterized by the simple addition reaction of HCl to acetylene in fixed bed reactors containing a catalyst of activated carbon impregnated with mercuric chloride. The reactors were shell and tube heat exchangers of carbon steel with catalyst packed in the tubes. Various techniques were used to ensure even distribution of flow through the many tubes, avoid hot spots, and obtain the maximum productivity from the catalyst. The rest of the process included feed gas drying, VCM recovery and purification, unreacted acetylene recovery, and waste disposal.

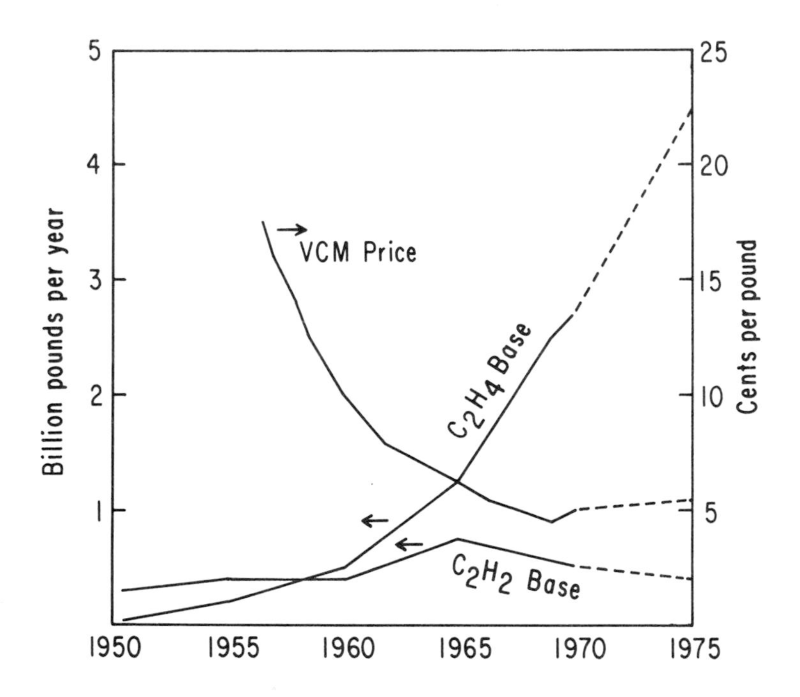

Figure 4. U.S. VCM production and price (data from Ref. 1)

The end product of the acetylene process was a high purity monomer that contained some water-derived impurities such as acetaldehyde but was free of some of the organic impurities such as butadiene, ethylene, and propylene which are often associated with other processes.

Acetylene-Ethylene Route. A "balanced" route (*8*), was developed simultaneously by several companies during the 1950's as cheaper petrochemical raw materials began entering the picture. In this process, the

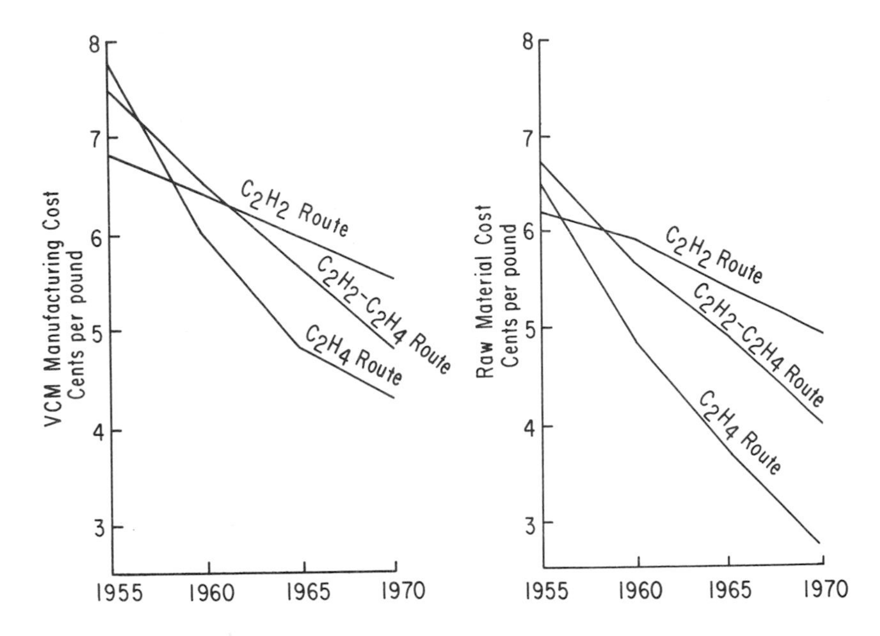

Figure 5. U.S. VCM manufacturing cost and raw material cost (based on data from Refs. 3, 6, and 7)

base was changed from all-acetylene to half acetylene and half ethylene, thus affording some relief from the relatively high price of acetylene. It also permitted a switch from anhydrous HCl to chlorine and consequently reduced the cost of the vinyl chloride materially.

This process is shown schematically in Figure 7. The ethylene part of the feed reacts with chlorine in the liquid phase to produce 1,2-dichloroethane (EDC) by a simple addition reaction, in the presence of a ferric chloride catalyst (*9*). Thermal dehydrochlorination, or cracking, of the intermediate EDC then produces the vinyl chloride monomer and by-product HCl (*1*). Acetylene is still needed as the other part of the over-all feed, to react with this by-product HCl and produce VCM as in the all-acetylene route.

The EDC cracking is an integral part of this process. Key points in the plant design include the cracking furnaces, which utilize alloy tubes such as Inconel or Incoloy, a quench tower for lowering process tempera-

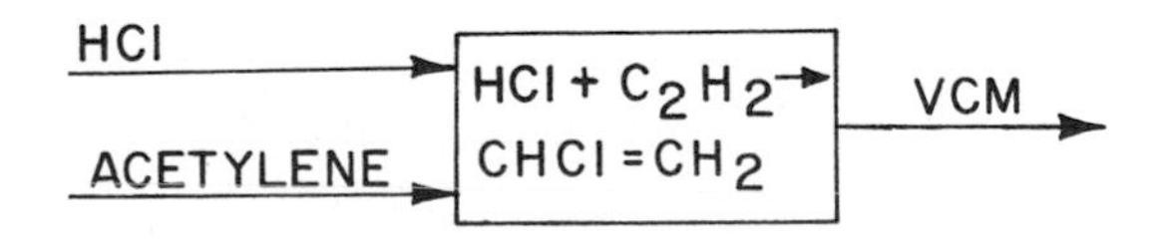

Figure 6. All-acetylene route to VCM

tures rapidly, and distillation columns for separating HCl and purifying VCM. The cracking pressure determines the design of the HCl column. With processes utilizing low pressures, the furnace exit gas must be compressed, or the HCl separator column must be designed for operation at extremely low temperature and high refrigeration load.

The EDC cracking processes that are being licensed most successfully are the joint process of Goodrich and Farbewerke Hoechst and those of Stauffer, Toyo Soda, and Kureha. These processes are generally similar, but the operating conditions can vary widely. Most plants use separate feed streams of acetylene and ethylene, but in some the feed streams are mixed. Union Carbide, Kureha, and Japanese Geon all have processes that produce VCM from mixed acetylene-ethylene feed stocks.

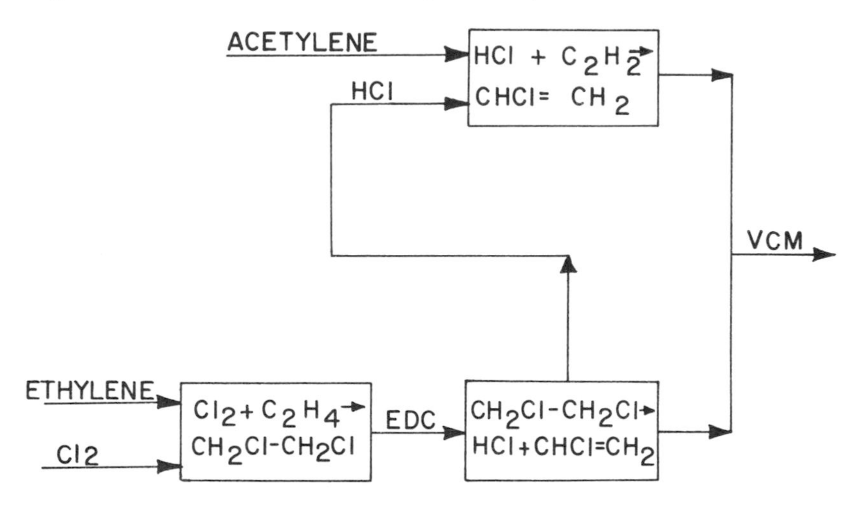

Figure 7. Balanced acetylene-ethylene route to VCM

The Wulff process, which is in operation in Europe and South America with varying degrees of success, uses twin regenerative furnaces for cracking light petroleum fractions. The mixed gas product from the furnace is first combined with HCl, which reacts with the acetylene to form VCM. Ethylene then reacts with chlorine to form EDC. An EDC cracking process converts EDC to VCM and HCl, and HCl is recycled to react with the acetylene in the mixed gas product.

The Kureha process, used in Japan, employs flame cracking of naphtha or, in some cases, heavier fractions. The treatment of the mixed gas is essentially the same as that described for the Wulff process.

In the Japanese Geon GPA process, aectylene in the mixed gas stream is solvent extracted and concentrated. The acetylene then reacts with HCl from an EDC cracking plant to produce VCM. The remaining

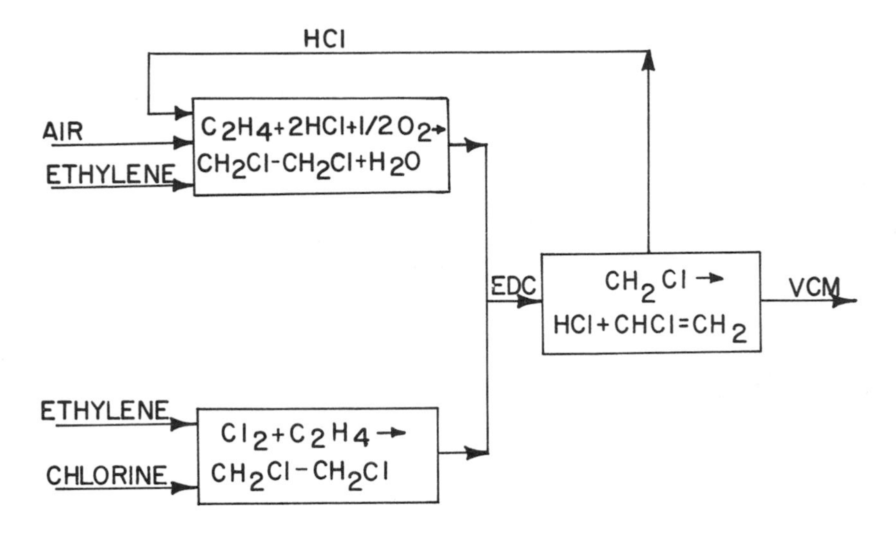

Figure 8. All-ethylene route to VCM

ethylene is chlorinated to EDC, which is then cracked to form VCM and HCl.

All-Ethylene Route. In the early 1960's an all-ethylene route (*6, 10, 11, 12, 13, 14, 15, 16, 17, 18*) to VCM, as shown in Figure 8, was made possible by development of the oxyhydrochlorination process. In this process, which replaced the acetylene-HCl part of the balanced route, the by-product HCl from EDC cracking reacts with ethylene and oxygen in the presence of a copper chloride catalyst to produce EDC.

The oxyhydrochlorination process can utilize either a fluid bed, a fixed bed, or fluid and fixed beds in series. Published economic data (3) indicate that these various reactor designs offer potentially the same economic returns; however, the fluid bed system has been the most successful in commercial operation because of the following:

(a) high one-pass conversion, eliminating raw material recycle streams

(b) superior temperature control, which produces a high purity crude product

(c) absence of hot spots, resulting in indefinite catalyst life

(d) ease of catalyst handling

(e) absence of reactor corrosion, permitting carbon steel construction

(f) high heat transfer coefficient, which allows direct generation of high pressure steam in the reactor.

The fixed bed system, which offers the same advantage of low raw material cost, also is licensed widely despite the possible disadvantages of reactor corrosion problems, low single-pass conversion, and poor temperature distribution resulting in hot spots. The fluid bed oxyhydrochlorination system is presently the most widely used in the industry.

Polymerization

The production of vinyl chloride monomer is only a part of PVC production. Polymerization of the monomer completes the process. Commercially, it is a batch operation by one of three methods: suspension, emulsion, or bulk. In all three methods, the chemical reaction is a free radical-initiated chain reaction. Peroxides or redox systems generally are used to provide the initial free radicals.

The suspension process, which was the first developed and is still the most common process used today, takes the approach of polymerizing in water to solve the difficult problem of heat removal. The monomer droplets are dispersed in water and maintained in suspension by agitation with the aid of a suspending agent such as poly(vinyl alcohol). Unreacted VCM is recovered, and the polymer is separated by centrifuging, dried, and packaged. The suspension process is used to produce general purpose resins that are later fabricated by extrusion, calendering, and injection molding.

Another method, the emulsion process, was developed in Europe to meet the demand for a resin with small particle size, high bulk density, and low plasticizer absorption properties. These characteristics are especially desirable in making plastisols and organosols, where the resins are dispersed in liquid plasticizers. In this polymerization process, emulsifiers are added, and the solution is agitated to keep the monomer droplets dispersed. The initiators must be water soluble; it is usually an inorganic persulfate or an organic hyperperoxide.

Unlike the suspension process where the reaction occurs in bulk within the monomer droplets, in the emulsion process it is initiated in the aqueous phase, and growth occurs within the monomer–polymer latex particles. Since the average particle size is too small to permit separation by centrifuging, the slurry is usually spray dried. When liquid plasticizers are mixed with the resulting resin, the product is a plastisol; addition of volatile solvents also makes an organosol. These liquid or paste products can then be used for dip coatings, rotational molding, slush molding, and also for spray and spread coatings.

A bulk process for polymerization has been historically difficult because of the problems of heat removal and particle agglomeration. Recently, a process has been developed by Pechiney-St. Gobain (*19*) where these problems are overcome by performing the main part of the reaction in a horizontal, water-cooled, autoclave using ribbon-type agitators. The vinyl chloride monomer and the initiators are charged to vertical, agitated, prepolymerizers where about 10% of the polymerization occurs. The charge is then transferred to the autoclaves where the reaction proceeds to about 25% completion. This process is now used widely; in the United States, Goodrich, Goodyear, and Hooker have such plants.

VCM Future

The current trend for vinyl chloride monomers is toward ethylene as the hydrocarbon raw material, replacing electrochemical acetylene, and it promises to continue. Electrochemical acetylene will be phased out almost completely, except in special cases. The high activation energy level required for forming the acetylene triple bond precludes design of a low cost process for its formation. In addition, the difficulties encountered in handling such a highly reactive material will deter its use.

The petrochemical acetylene processes using light petroleum fractions, such as the Wulff, Kureha, and Japanese Geon processes are interesting in a strict economic sense, but according to a published report (*20*), one of these encountered technical problems in operation. It is believed that the problem was in maintaining the desired 1:1 ratio of acetylene to ethylene (*21*).

Other processes that may come to the fore soon are those for producing dilute ethylene, such as the Diamond-de Nora Dianor process (*22*). Kureha and Lummus have similar processes. In these, simplification of the ethylene purification step allows significant reduction in investment cost, refrigeration cost, and compression cost. The cheaper ethylene can then be incorporated into ethylene-route processes for VCM to decided advantage, especially for smaller units.

Single-step processes for ethylene to VCM (*3*) and ethane to VCM (*23*) are under review or development by some companies, but the dissimilar conditions of the separate reactions make it unlikely that commercial processes will result within the next five years.

PVC Future

The already low cost of converting VCM to PVC probably precludes any revolutionary new process replacing the current suspension and emulsion processes. The Pechiney-St. Gobain bulk polymerization process is being tested thoroughly; within the next year it will be evident if the process offers advantages, either in product quality or conversion cost.

A significant forward step that could be taken in the polymerization operation would be the development of a continuous process rather than the batch processes now available. It is not expected that continuous polymerization will be commercially feasible within the next few years.

Potential PVC process improvements that could contribute to significant economic advantage include:

(a) faster reaction rates by using catalyst mixtures to produce a linear reaction rate with time

(b) larger polymerizers that would result in lower investment costs and higher quality resins

(c) reduced turnaround time and manpower requirements by using such equipment as automatic hydraulic reactor cleaners.

The most significant progress in the near future will be in the development of specialized resins for use in specific end products which will exploit more fully the full versatility and utility of PVC resin and its end consumer products. No significant changes are expected in the present methods of production of VCM and polymerization to PVC, but production rates and efficiencies are expected to continue to rise during the next five years.

Literature Cited

(1) Caudle, P. G., *Chem. Ind.* (1968) 1551.
(2) U. S. Tariff Commission Report (1969).
(3) Onoue, Y., Sakurayama, K., *Chem. Econ. Eng. Rev.* (Nov. 1969) 17.
(4) Badger-Tomoe, personal communication, October 4, 1967.
(5) *Modern Plastics* (1970) **47** (1), 73.
(6) Buckley, J. A., *Chem. Eng.* (Nov. 21, 1966) 102.
(7) Rosenzweig, M. D., *Chem. Eng.* (Nov. 3, 1969) 64.
(8) Albright, L. F., *Chem. Eng.* (March 27, 1967) 123.
(9) Distillers Co., British Patent **553,959** (1943).
(10) *Hydrocarbon Processing* (1967) **46** (11), 239.
(11) Gatto, J. D., *Rubber World* (1965) **152** (6), 81.
(12) *Chem. Week* (1964) **95** (9), 101.
(13) Albright, L. F., *Chem. Eng.* (April 10, 1967) 219.
(14) Edwards, E. F., Weaver, T., *Chem. Eng. Progr.* (1965) **61** (1), 21.
(15) Burke, D. P., Miller, R., *Chem. Week* (1964) **95** (8), 93.
(16) *Japan Chem. Week* (Sept. 25, 1969) 4.
(17) *Chem. Eng.* (Nov. 8, 1965) 245.
(18) Geldart, D., *Chem. Ind.* (1967) 1474.
(19) Thomas, J. C., *Hydrocarbon Processing* (1968) **47** (11), 192.
(20) *Chem. Eng.* (Nov. 18, 1968) 87.
(21) *Chem. Week* (1970) **106** (15), 61.
(22) Remirez, R., *Chem. Eng.* (April 22, 1968) 142.
(23) Monsanto Co., British Patent **979,001** (1965).

RECEIVED May 14, 1970.

11

Homogeneous Catalysis: Progress, Problems, and Prospects

ERIC W. STERN

Engelhard Industries, 497 Delancy St., Newark, N. J. 07105

Research in catalysis by coordination complexes has resulted in the discovery of many new reactions and the development of several important industrial processes. Close parallels to heterogeneous and biocatalysis are indicated. Progress in reaching a detailed understanding of homogeneous catalytic reactions has been slower. Problems encountered are that nonisolable catalytic species may form in situ, *kinetic studies often provide only limited mechanistic information, observed rates and products are affected strongly by media and catalyst compositions, and analogies between apparently closely related systems are frequently invalid. Difficulties in commercialization include containment of corrosive systems and product isolation from liquid media. These problems appear resolvable by currently available means. It is expected that continued research will lead to new applications and will contribute substantially to improved understanding of catalytic phenomena in general.*

During the past decade homogeneous catalysis has progressed from an interesting novelty to its present status as a recognized field. In the interval a great deal has been said and written concerning the subject, much of it well beyond the scope of this discussion which is limited to an indication of what has been accomplished, an outline of the problems which have been encountered, and a forecast of future prospects.

In the strictest sense, homogeneous catalysis involves catalytic reactions occurring in a single phase. However, as currently used, the term implies only that at least a portion of a particular reaction is known or suspected strongly to occur in the coordination sphere of a metal (most frequently a transition metal). Activation of substrates and likely the steric course of the reaction are then consequences of bonding in an intermediate complex. The definition is not necessarily limited to reactions which are homogeneous in the physicochemical sense. While most reactions included in this category are carried out in the liquid phase, slurry, gas–liquid, and even gas–solid systems are not excluded. Homogeneous reactions, such as general acid–base catalysis or radical chain processes, usually are not considered unless metal complexes play a part. Thus, designation of the area as coordination catalysis would probably be more descriptive.

Progress

The major accomplishment of research in homogeneous catalysis to date has been the discovery of an impressive number of new reactions. Several of these have been developed into industrially important processes. In addition, the list of complexes which have potential as new catalysts or as models for reaction intermediates grows steadily. While the incentive for discovering new reactions and seeking practical applications for these remains understandably high, increasingly serious efforts are being made to gain insight into intimate reaction detail.

The following is an attempt to illustrate the variety of results obtained. Obviously, any of the topics cited could provide the basis for extensive discussion.

Reactions. OXIDATIONS. *Nucleophilic Reactions of Olefins.* These reactions, which over-all, involve replacement of vinyllic hydride by a nucleophile (Reaction 1),

$$RCH{=}CH_2 + HX \xrightarrow[\substack{Cu(II)\\O_2}]{Pd(II)} \underset{\substack{|\\X}}{RC}{=}CH_2 + RCH{=}CH{-}X \qquad (1)$$

are, at present, probably the best known and most widely studied homogeneous catalytic reactions. Indeed, the current high interest in this area of chemistry can be traced to the disclosure of the now familiar Wacker synthesis of acetaldehyde from ethylene and water by this route. The reaction which converts olefins to aldehydes and/or ketones in aqueous media (*1*), yields vinyl compounds in nonaqueous systems and has been demonstrated with such nucleophiles as alcohols (*2, 3*), carboxylic acids and their salts (*2, 3*), amines and amides (*3*), cyanide (*4*), and carbanions (*5*). Like many homogeneously catalyzed oxidations, it consists actually of two separate stoichiometric reactions, in this instance, nucleophilic attack on olefin leading to reduction of the "catalyst" (Reaction 2) and re-oxidation of the metal to its active state (Reaction 3).

$$RCH{=}CH_2 + PdX_2 + HX \rightarrow RCH{=}CH{-}X + Pd(0) + 2HX \qquad (2)$$

$$Pd(0) + 2CuX_2 \rightarrow PdX_2 + 2CuX \qquad (3)$$

$$2CuX + 2HX + 1/2O_2 \rightarrow CuX_2 + H_2O$$

The over-all process is catalytic both in metal and regenerating agent.

While this type of reaction has been demonstrated with most of the group VIII noble metals, palladium has the highest activity and, therefore, is normally used. Regenerating agents (co-catalysts) are most frequently $CuCl_2$, $FeCl_3$, and quinone. Reaction details are complex (*6*).

Oxidation of Alcohols. Conversion of alcohols to aldehydes and acids has received considerably less attention. The chemistry of this process has been explored primarily with respect to formation of stable hydrides of Pt, Ir, Ru, and Os (*7, 8*) (Reaction 4).

$$(Et_3P)PtCl_2 + EtOH + KOH \rightarrow (Et_3P)_2PtHCl + CH_3CHO + KCl + H_2O \qquad (4)$$

When the metal hydride formed is unstable (as in the case of Pd), the reaction can be used as the basis for a catalytic conversion of alcohols (*9, 10*) (Reaction 5).

$$RCH_2OH + 1/2O_2 \xrightarrow[20\%\ H_2SO_4]{PdSO_4} RCHO + H_2O \qquad (5)$$

Oxidative Coupling. A number of cases of this reaction type have been reported. Again, all are really two-step processes involving reduction of metal in the coupling step followed by *in situ* re-oxidation. Recent examples are coupling of aromatics (*11*) (Reaction 6),

$$2C_6H_6 + 1/2O_2 \xrightarrow[HOAc,\ HClO_4]{Pd(OAc)_2} C_6H_5 - C_6H_5 + H_2O \qquad (6)$$

aromatics and olefins (*12*) (Reaction 7),

$$\phi CH{=}CH_2 + \phi H + Pd(OAc)_2 \rightarrow \phi CH{=}CH\phi + Pd(0) + 2HOAc \qquad (7)$$

$$CH_2{=}CHOAc \xrightarrow[Cu(OAc)_2]{Pd(OAc)_2} AcO{-}CH{=}CH{-}CH{=}CH{-}OAc \qquad (8)$$

$$\phi{-}\underset{\displaystyle CH_3}{\underset{|}{C}}{=}CH_2 \rightarrow \phi{-}\underset{\displaystyle CH_3}{\underset{|}{C}}{=}CH{-}CH{-}CH{=}\underset{\displaystyle CH_3}{\underset{|}{C}}{-}\phi \qquad (9)$$

and β-substituted α-olefins (*14, 15*) (Reactions 8 and 9).

Older examples are coupling of phenols catalyzed by Cu(I)–pyridine in which the relative amounts of carbon-carbon and carbon-oxygen coupling are controlled by the copper/pyridine ratio (*16*) (Reaction 10)

$$ArOH \xrightarrow[Py,\ O_2]{Cu(I)} HO-C_6H_4-C_6H_4-OH \quad (10)$$

$$\searrow \left(-C_6H_4-O-C_6H_4-O-\right)_n$$

and coupling of thiols to disulfides (*17*). While many aspects of oxidative coupling resemble radical processes, it is by no means clear that radical species are actually involved (*18*).

Acetoxylation of Aromatics. Reaction 11,

$$\phi H + HOAc + PdX_2 \rightarrow \phi OAc + Pd(0) + 2HX \quad (11)$$

originally found as a side reaction in aromatic coupling, can be made highly selective by a proper choice of conditions. It has been examined both with respect to phenyl acetate and benzyl acetate formation (*19, 20*).

Autoxidation. In this category, which refers to direct reactions of substrates with oxygen, there are numerous examples of metal catalysis of hydrocarbon oxidation. While metal participation most frequently involves catalysis of hydroperoxide decomposition (*21, 22*) (Reactions 12 and 13),

$$ROOH + M^{+n} \rightarrow RO_2\cdot + M^{+(n-1)} + H^+ \quad (12)$$

$$ROOH + M^{+(n-1)} \rightarrow RO\cdot + M^{+n} + OH^-$$

participation in initiation (*23–26*) and termination (*27*) steps has also been found. In addition, such recently discovered cases as catalysis of the conversion of phosphines to phosphine oxides (Reaction 13)

$$Ar_3P + 1/2O_2 \xrightarrow{(Ar_3P)_4Pd} Ar_3PO \quad (13)$$

and of isocyanides to isocyanates (*28–30*) as well as oxidations of SO_2 (*31*), CO (*32*), NO, and NO_2 (*33*) should be mentioned.

REDUCTIONS. *Hydrogenation of Mono- and Polyolefins and Acetylenes.* Catalysis of hydrogenation of unsaturated hydrocarbons has been studied widely. Catalysis by a variety of metal complexes including those of Pt, Pd, Rh, Ru, Ir, Os, Cr, and Co has been demonstrated (*34*). Particular attention has been given to the catalytic properties of $(\phi_3P)_3RhCl$, the so-called Wilkinson catalyst (*35–37*), Pt(II)–$SnCl_2$ complexes (*38*), and $[Co(CN)_5]^{3-}$ (*39, 40*). Many detailed studies have been carried out, and some comparisons with heterogeneous hydrogenation catalysis have been made (*41, 42*).

Hydrogenation of Functional Groups. In most instances homogeneously catalyzed hydrogenation of olefins and acetylenes takes place without affecting other functional groups. However hydrogenation of various organic functions has been reported. Among these are ketones (*43*), diazo compounds (*44*), imines (*44*), nitro compounds (*44*), and sulfoxides (*45*).

Reductive Coupling. Only one example of this interesting reaction has been disclosed (*46*). This is the coupling of vinyl chloride to butadiene (Reaction 14).

$$CH_2{=}CH{-}Cl + 2SnCl_2 \xrightarrow[\substack{CsF_2 \\ DMF,\ H_2O}]{Pt(II)} CH_2{=}CH{-}CH{=}CH_2 + 2SnCl_3 \quad (14)$$

The role of the CsF_2 co-catalyst is unclear.

ISOMERIZATION. *Double Bond.* The isomerization of olefinic double bonds frequently accompanies hydrogenations (*47, 48*). In addition, the reaction is promoted by such complexes as $RhCl_3$ (*49*), $(\phi_3P_2RuCl_2$ (*50*), $(\phi CN)PdCl_2$ (*51*), and $[(C_2H_4)PtCl]_2$ (*52*).

VALENCE. Several examples of these thermally forbidden processes, exemplified by the conversion of quadricyclene to norbornadiene (Reaction 15)

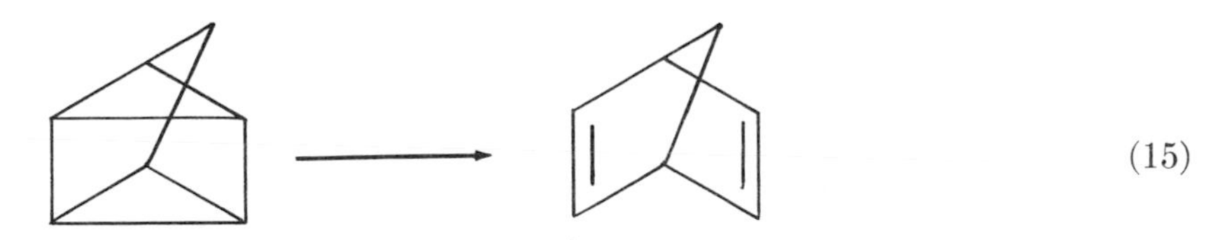

(15)

have been reported (*53–57*). Catalysts include $PdCl_2$, Pd(COD)Cl_2, π-allyl PdCl, norbornadiene-rhodium and platinum chloride complexes, and $[(C_2H_4)RhCl]_2$.

CARBONYLATION–DECARBONYLATION. These reactions constitute a large and important segment of the homogeneous catalytic literature. Catalysis by most of the group VIII noble metals has been found, and a variety of substrates have been converted. For example, olefins react with CO and $PdCl_2$ or $(\phi_3P)_2PdCl_2$ to yield β-chloroacyl chlorides (*58*) and unsaturated acyl chlorides (*59*), respectively. α, ω-Dienes, such as

1,5-hexadiene, yield cyclic ketoesters in the presence of Pd(II) in alcoholic media (*60*) (Reaction 16).

$$\text{1,5-hexadiene} + CO \xrightarrow[\substack{CH_3OH \\ HOTs}]{(Bu_3P)_2PdI_2} \text{(2-methyl-5-(}CH_2COOCH_3\text{)cyclopentanone)} \quad (16)$$

Amines are converted to isocyanates in the presence of $PdCl_2$ (*61*), and methanol is carbonylated to acetic acid (*62*) (Reaction 17).

$$CH_3OH + CO \xrightarrow[\substack{CH_3I \\ \phi H}]{(\phi_3P)_2RhCOCl} CH_3COOH \quad (17)$$

The hydroformylation of olefins (Reaction 18)

$$R{-}CH{=}CH_2 + CO + H_2 \rightarrow RCH_2CH_2CHO + \underset{\substack{| \\ CHO}}{RCH}{-}CH_3 \quad (18)$$

must also be included in this category. This can be carried out under relatively mild conditions with such catalysts as $(\phi_3P)_2RhCOCl$ (*63*) and various phosphine complexes of cobalt (*64*).

While most of the above carbonylations are carried out at pressures greater than 40 atm (isocyanate and acetic acid formations are exceptions), decarbonylations are low pressure reactions. Decarbonylation of acyl halides catalyzed by $(\phi_3P)_2RhCOCl$ leads either to halides (*65*) (Reaction 19)

$$\phi COI \rightarrow \phi I + CO \quad (19)$$

or olefins (*66*) (Reaction 20),

$$RCH_2CH_2COCl \rightarrow RCH{=}CH_2 + CO + HCl \quad (20)$$

depending on the presence of β-hydrogens. Similarly, aldehydes have been decarbonylated (*67*) to paraffins. Both addition and removal of CO are thought to proceed through common intermediates (*68*).

ADDITION REACTIONS. This category includes polymerizations catalyzed by a number of soluble Ziegler-Natta type materials as well as various oligomerizations. The latter includes conversion of ethylene to butenes catalyzed by rhodium and ruthenium chlorides (*69*) and $[(C_2H_4)\text{-}PdCl_2]_2$ (*70*), the codimerization of ethylene and butadiene in which almost no dimer of either component is formed catalyzed by $RhCl_3$ (*69, 71*) and $CoCl_2$-di-*tert*-phosphine–$AlEt_3$ (*73*) (Reaction 21),

$$C_2H_4 + C_4H_6 \rightarrow CH_2{=}CH{-}CH_2{-}CH{=}CH{-}CH_3 + CH_3{-}CH{=}CH{-}CH{=}CH{-}CH_3 \quad (21)$$

cyclooligomerization of acetylenes (*73, 74*) (Reaction 22)

$$3\phi C{\equiv}C\phi \xrightarrow{(\phi CN)_2PdCl_2} C_6\phi_6 \quad (22)$$

and the telomerization of butadiene dimerization catalyzed by Pd(II) and other group VIII noble metal salts and complexes (*75*) (Reaction 23).

$$\phi OH + 2CH_2{=}CH{-}CH{=}CH_2 \rightarrow \phi O{-}CH_2{-}CH{=}CH{-}CH_2{-}CH_2CH_2{-}CH{=}CH_2 \quad (23)$$

DISPLACEMENTS. Most interesting among these reactions is the replacement of unactivated vinyl halides by such nucleophiles as acids (*76*), their salts (*77*), alcohols (*76*), and other halides (*77*) under extremely mild conditions (Reaction 24).

$$CH_2{=}CH{-}X + HY \xrightarrow[25°C]{Pd(II)} CH_2{=}CH{-}Y + HX \quad (24)$$

The reaction appears to be stereospecific (*78, 79*). Related reactions are the trans-esterification of vinyl esters (*80*) (Reaction 25)

$$CH_2CHOCOR + R'COOH \xrightarrow{Pd(II)} CH_2CHOCOR' + RCOOH \quad (25)$$

and the conversion of vinyl esters to vinyl ethers (*81*) (Reaction 26).

$$CH_2{=}CHOCOR + R'OH \rightarrow CH_2{=}CHOR' + RCOOH \quad (26)$$

DISPROPORTIONATIONS. The disproportionation of olefins to higher and lower homologs (Reaction 27)

$$2RCH{=}CHR' \rightleftarrows RCH{-}CHR + R'CH{-}CHR' \quad (27)$$

which has been investigated primarily as a heterogeneously catalyzed reaction (*82*) has also been found to occur in the presence of such soluble complexes as $(\phi_3P)_2Mo(NO)_2Cl_2$ and its tungsten analog (*83*) and a WCl_6-$EtAlCl_2$-EtOH system (*84*). Also, disproportionation of 1,4-cyclohexadiene to benzene and cyclohexene has been catalyzed by the Vaska complex, $(\phi_3P)_2IrCOCl$ (*85*).

Industrial Applications. Several large scale industrial processes are based on some of the reactions listed above, and more are under development. Most notable among those currently in use is the already mentioned Wacker process for acetaldehyde production. Similarly, the production of vinyl acetate from ethylene and acetic acid has been commercialized. Major processes nearing commercialization are hydroformylations catalyzed by phosphine–cobalt or phosphine–rhodium complexes and the carbonylation of methanol to acetic acid catalyzed by $(\phi_3P)_2RhCOCl$.

Complexes of Potential Interest. Research in homogeneous catalysis has been aided greatly by the preparation of many new complexes which may serve either as catalysts or as models for reaction intermediates. Among those of particular interest are olefin and acetylene complexes, alkyls, hydrides, carbonyls, and complexes of molecular oxygen, nitrogen, and SO_2. Many of these materials bear a striking resemblance to species thought to form on the surfaces of heterogeneous catalysts, and study of their reaction chemistry is expected to shed new light on established catalytic reactions and lead to the discovery of new reactions. To date such studies have led to the recognition of a number of reaction types—*e.g.*, oxidative addition, reductive elimination, and insertion—as well as conditions favorable for "activation" of substrates—*e.g.*, coordinative unsaturation, that is, the occurrence of a vacant or weakly bound coordination site.

Since many complexes resemble proposed intermediates in heterogeneous and biocatalysis and because their chemistry resembles that found in these less readily accessible systems, model studies involving relatively simple complexes are becoming more frequent in these areas. While such studies are, as yet, in their infancy, they have refocused attention on the chemistry of the catalytic systems under investigation. For example, evidence for the formation of π-olefin and π-allyl complexes as chemisorbed species is being found (*86*), and it is no longer unusual to hear reactive centers described in the language of coordination chemistry. Model studies have been carried out in hydrogenation, isomerization, and H–D exchange processes in which the chemistry of complexes is being compared with that of the same metals in heterogeneous catalytic systems (*42, 87, 88*). Valuable information and support for the validity of such modeling has been obtained in studies in which the chemistry of vitamin B_{12} is mimicked by simple bis(dimethylglyoximato)Co(II) complexes (*89*). Finally, complexes of molecular nitrogen are being studied intensely with the dual aim of achieving a low pressure ammonia synthesis and of elucidating the mechanism of enzymatic nitrogen fixation (*90, 91*).

Problems

While considerable progress has been made, problems both in interpreting and in applying research results have arisen.

Interpretational Problems. Despite considerable research efforts, the intimate mechanisms of most homogeneous catalytic reactions are still poorly understood. Perhaps because much discussion has centered around the classification of reaction steps (*e.g.*, insertion, oxidative addition, etc.) and various attempts to correlate these with bonding in particular oxidation states and the recognition of conditions favorable for complex formation (coordinative unsaturation), the feeling has developed that what is known about homogeneous catalysis is reasonably well understood in terms of reactive species, intermediates, and fundamental steps. However, this descriptive language is not very different from discussing heterogeneous catalysis in terms of chemisorption, rearrangement of surface species, and desorption without specifying the detailed nature of these steps or of assuming special catalytic properties for particular crystal faces or defects. Moreover, the various empirical schemes suggested thus far tend to become limited in applicability as more data become available. The problems encountered are common to catalytic research as a whole and are summarized in the discussion below.

In Situ FORMATION OF ACTUAL CATALYTIC SPECIES. One of the original hopes of workers in homogeneous catalysis is that the catalytic species would be the salt or complex added to the system; this hope has not been realized. It is now generally recognized that in most cases, actual catalytic materials are formed from the components of particular systems. In many instances, the precise nature of such species remains elusive and a subject for continuing discussion. For example, in isomerizations catalyzed by rhodium chloride, the actual catalyst is thought to be a Rh(I) species rather than the Rh(III) originally added (*92, 93*). Similarly, in the dimerization of ethylene catalyzed by Pd complexes, catalysis by a Pd–C_2H_4 complex is implicated (*70*).

KINETIC STUDIES PROVIDE ONLY LIMITED MECHANISTIC INFORMATION. While such studies are invaluable and frequently indicate the nature of pre-rate-determining steps, they provide almost no information concerning such vital fast steps as electron transfers and rearrangements. For example, despite extensive studies of the kinetics of acetaldehyde and vinyl acetate syntheses, it is clear only that olefin, nucleophile, and palladium combine in a complex. The nature of the rate-determining step as well as the details of post-rate determining product forming steps remains uncertain (*7, 94*). In some cases—*e.g.*, the metal-catalyzed autoxidation of thiols to disulfides—re-oxidation of metal to its catalytically active state is rate determining, a result which provides no insight into the nature of the product forming reaction (*95, 96*).

OBSERVED RATES AND PRODUCTS DEPEND STRONGLY ON MEDIA AND CATALYST COMPOSITIONS. It is generally accepted that catalysis by metals in solution is not catalysis by an ion in a particular oxidation state but by the ion with its full complement of ligands and possibly its solvation sphere as well. That is, the composition and configuration of both inner and outer coordination spheres frequently determine the course of a particular reaction. While this may cause difficulties in sorting out varying results obtained under apparently similar conditions, it provides the largest hope for realizing one of the primary aims of catalytic research—the design of a catalyst to perform a specific function.

Many examples of this problem exist, and only a few are cited here. For instance, in the formation of vinyl acetate, no 1,2-disubstituted saturated products are found in the absence of Cu(II), originally thought to be merely a regenerating agent for palladium (*7*). In Cu(II)-containing systems, at high chloride concentrations, the formation of vinyl and 1,1-disubstituted products is almost completely suppressed, and products are almost entirely 1,2 (*97*). Likewise, in the reaction of propylene with acetic acid, the ratio of primary to secondary vinyl esters is greatly increased in 1,2-dimethoxyethane over that found in isooctane (*7*). Similar effects have been noted by adding benzonitrile to acetic acid solution (*98*), while DMSO promotes allyl ester formation (*99*). Increasing acetate and chloride concentrations relative to Pd lead to increasing terminal substitution (*100*), while the presence of nitrates promotes 1,2-disubstitution (*101, 102*).

In the reaction of toluene to form bitolyls and benzyl acetate catalyzed by $PdCl_2$ in the presence of NaOAc, the bitolyl/benzyl acetate ratio has been found to vary from 3.8 to 0.0074 as the sodium acetate/$PdCl_2$ ratio is varied from 50–20 (*103*).

Finally, in the presence of $Pd(OAc)_2$ in chloride free systems vinyl acetate is coupled to 1,4-diacetoxybutadiene (*15*) (Equation 8), while in the presence of $PdCl_2$/NaOAc mixtures it decomposes to acetaldehyde and acetic anhydride (*104*) (Reaction 28).

$$CH_2{=}CHOAc + HOAc \xrightarrow[NaOAc]{PdCl_2} CH_3CHO + Ac_2O \tag{28}$$

ANALOGIES BETWEEN APPARENTLY CLOSELY RELATED SYSTEMS ARE FREQUENTLY INVALID. This is a bothersome problem since often it is desirable to find systems which may be more amenable to study—*e.g.*, those in which intermediate complexes are more stable. However, while it was originally thought that what was learned from the behavior of one metal in a particular subgroup in the periodic table would be valid for another member of the same subgroup and while similar reactions are found frequently, enough exceptions exist to make the value of such analogies doubtful. For example, C_2H_4–$PtCl_2$ decomposes thermally to yield chiefly chlorinated C_2's (*105*), while the corresponding Pd complex yields butenes (*10*). In the catalysis of olefin hydrogenation, $(\phi_3P)_2$-$Pt(CN)_2$–$SnCl_2$ is ineffective while the corresponding Pd complex is active. Conversely, $(\phi_3As)_2PtCl_2$–$SnCl_2$ is effective while the corresponding Pd complex is ineffective (*48*).

Industrial Problems. Problems have also been encountered in attempts to commercialize various homogeneous catalytic reactions. These, in addition to the highly corrosive nature of many metal solution systems involve general problems of large scale handling of liquid systems such as mass and heat transfer and the isolation of products from solution. Moreover, the recovery and/or regeneration of metals often presents difficulties.

Prospects

While these problems exist at present, the situation is certainly far from hopeless. As far as interpretational difficulties are concerned, the recognition of various complexities in the reactions studied is a spur to finding answers to the questions raised. At any rate, currently available investigative techniques are by no means exhausted. Considerable expansion of understanding can be expected as reactions are studied in greater detail.

Industrial problems have, in some instances, been solved either by a proper choice of construction materials and suitable process design or by development of heterogeneous catalytic systems using supported complexes or by generating active complexes *in situ* on a support material which avoid some of the problems of liquid-phase operation. For example, a number of the problems in liquid-phase vinyl acetate processing have been overcome by development of supported Pd catalysts (*106*). Vapor-phase hydroformylation has been carried out on supported rhodium complexes (*107*).

In general, the future for homogeneous catalysis appears bright. By some estimates (*108*), in 12–20 years some 50% of heavy organic chemicals will be produced by complex derived processes. There are inherent advantages to homogeneous catalysis in that frequently it provides pathways for reactions which are either extremely difficult or impossible to carry out by other means, in many instances allowing the use of cheaper raw materials. Because such reactions often occur under mild conditions, they tend to yield desirable products in high selectivity. Homogeneous catalysts are also potentially more efficient than heterogeneous catalysts in that a reacting substrate presumably has ready access to each molecule of catalyst metal rather than to suitably disposed surface sites only. Thus, we may expect both lower raw material and plant investment costs as a result of more widespread use of homogeneous catalysts. Finally, there is continuing hope that homogeneous catalysis will offer new insights into heterogeneous and biocatalysis, leading to equally profitable advances in these areas.

Literature Cited

(1) Smidt, J., Hafner, W., Jira, R., Sedlmeier, J., Sieber, R., Ruttinger, R., Kojer, H., *Angew. Chem.* (1959) **71,** 176.
(2) Moiseev, I. I., Vargaftik, M. N., Syrkin, Ya. K., *Dokl. Akad. Nauk, SSSR* (1960) **133,** 377.
(3) Stern, E. W., Spector, M. L., *Proc. Chem. Soc.* (1961) 370.
(4) Odaira, Y., Oishi, T., Yukawa, T., Tsutsumi, S., *J. Am. Chem. Soc.* (1966) **88,** 4105.
(5) Tsuji, J., Takahashi, H., *J. Am. Chem. Soc.* (1965) **87,** 3275.
(6) Stern, E. W., *Catalysis Rev.* (1967) **1,** 73.
(7) Chatt, J., Shaw, B. L., *Chem. Ind.* (1960) 931; (1961) 290.
(8) Vaska, L., Di Luzio, J. W., *J. Am. Chem. Soc.* (1961) **83,** 1262, 2784.
(9) Brown, R. G., Davidson, J. M., Triggs, C., *Preprints, Div. of Petrol. Chem., Am. Chem. Soc.* (1969) **14** (2), B23.
(10) Lloyd, W. G., *J. Org. Chem.* (1967) **32,** 2816.
(11) Davidson, J. M., Triggs, C., *J. Chem. Soc. (A)* (1968) 1324.
(12) Fujiwara, Y., Moritani, I., Matsuda, M., Teranishi, S., *Tetrahedron Letters* (1968) 3863.
(13) Uemura, S., Okada, T., Ichikawa, K., *Nippon Kagaku Zasshi* (1968) **89,** 692.
(14) Volger, H. C., *Rec. Trav. Chim.* (1967) **86,** 677.
(15) Kohll, C. F., Van Helden, R., *Rec. Trav. Chim.* (1967) **86,** 193.
(16) Endres, G. F., Hay, A. S., Eustance, J. W., *J. Org. Chem.* (1963) **28,** 1300.
(17) Cullis, C. F., Trim, D. L., *Discussions Faraday Soc.* (1968) **46,** 144.
(18) Stern, E. W., "Homogeneous Catalysis," G. N. Schrauzer, Ed., Marcel Decker, New York, in press.
(19) Davidson, J. M., Triggs, C., *J. Chem. Soc. (A)* (1968) 1331.
(20) Bryant, D. R., McKeon, J. E., Ream, B. C., *J. Org. Chem.* (1968) **33,** 4123.
(21) Kamiya, Y., Beaton, S., Lafortune, A., Ingold, K. U., *Can. J. Chem.* (1963) **41,** 2020.
(22) Copping, C., Uri, N., *Discussions Faraday Soc.* (1969) **46,** 202.
(23) Cooper, T. A., Clifford, A. A., Mills, D. J., Waters, W. A., *J. Chem. Soc. (B)* (1966) 796.
(24) Uri, N., *Nature* (1956) **177,** 1177.
(25) Stern, E. W., *Chem. Commun.* (1970) 736.
(26) Heiba, E. I., Dessau, R. M., Koehl, Jr., W. J., *J. Am. Chem. Soc.* (1968) **90,** 1082.
(27) Bacha, J. D., Kochi, J. K., *J. Org. Chem.* (1968) **33,** 83.
(28) Takahashi, S., Sonogashira, K., Hagihara, N., *Mem. Inst. Sci. Ind. Res., Osaka Univ.* (1966) **23,** 69.
(29) Wilke, G., Schott, H., Heimbach, P., *Angew. Chem., Intern. Ed.* (1967) **6,** 92.

(30) Cook, C. D., Jauhal, G. S., *Inorg. Nucl. Chem. Letters* (1967) **3**, 31.
(31) Cook, C. D., Jauhal, G. S., *J. Am. Chem. Soc.* (1967) **89**, 3066.
(32) Nyman, C. J., Wymore, C. E., Wilkinson, G., *J. Chem. Soc. (A)* (1968) 561.
(33) Collman, J. P., Kubota, M., Hosking, J. W., *J. Am. Chem. Soc.* (1967) **89**, 4809.
(34) Lyons, J. E., Rennick, L. E., Burmeister, J. L., *Ind. Eng. Chem., Prod. Res. Develop.* (1970) **9**, 2.
(35) Osborn, J. A., Jardine, F. H., Young, J. F., Wilkinson, G., *J. Chem. Soc. (A)* (1966) 1711.
(36) Montelatici, S., vander Ent, A., Osborn, J. A., Wilkinson, G., *J. Chem. Soc. (A)* (1968) 1054.
(37) Hussey, A. S., Takeuchi, Y., *J. Org. Chem.* (1970) **35**, 643.
(38) Adams, R. W., Bateley, G. E., Bailar, Jr., J. C., *J. Am. Chem. Soc.* (1968) **90**, 6051.
(39) Kwiatek, J., *Catalysis Rev.* (1967) **1**, 37.
(40) Pregaglia, G., Morelli, D., Conti, F., Gregorio, G., Ugo, R., *Discussions Faraday Soc.* (1968) **46**, 110.
(41) Jardine, I., Howsam, R. W., McQuillin, J., *J. Chem. Soc. (C)* (1969) 260.
(42) Smith, G. V., Shuford, R. J., *Tetrahedron Letters* (1970) 525.
(43) Schrock, R. R., Osborn, J. A., *Chem. Commun.* (1970) 567.
(44) Jardine, I., McQuillin, F. J., *Chem. Commun.* (1970) 626.
(45) James, R., Ng, F. T. T., Rempel, G. L., *Can. J. Chem.* (1969) **47**, 452.
(46) Jones, F. N., *J. Org. Chem.* (1967) **32**, 1667.
(47) Horner, L., Buthe, H., Siegel, H., *Tetrahedron Letters* (1968) 4023.
(48) Bailar, J. C., Itatani, H., *J. Am. Chem. Soc.* (1967) **89**, 1592.
(49) Harrod, J. F., Chalk, A. J., *J. Am. Chem. Soc.* (1966) **88**, 3491.
(50) Blum, J., Pickholtz, Y., *Israel J. Chem.* (1969) **7**, 723.
(51) Sparke, M. B., Turner, L., Wenham, A. J. M., *J. Catalysis* (1965) **4**, 332.
(52) Harrod, J. F., Chalk, A. J., *J. Am. Chem. Soc.* (1964) **86**, 1776.
(53) Hogeveen, H., Volger, H. C., *J. Am. Chem. Soc.* (1967) **89**, 2486.
(54) Katz, T. J., Cerefice, S. A., *Tetrahedron Letters* (1969) 2561.
(55) Volger, H. C., Hogeveen, H., Gassbeek, M. M. P., *J. Am. Chem. Soc.* (1969) **91**, 218.
(56) Frye, H., Kuljian, E., Viebrock, J., *Inorg. Nucl. Chem. Letters* (1966) **2**, 119.
(57) Trebellas, J. C., Olechowski, J. R., Jonassen, H. B., *J. Organometal. Chem.* (1966) **6**, 412.
(58) Tsuji, J., Morikawa, M., Kiji, J., *J. Am. Chem. Soc.* (1964) **86**, 4851.
(59) Bittler, K., Kutepow, N. V., Neubauer, D., Reis, H., *Angew. Chem., Intern. Ed.* (1968) **7**, 329.
(60) Brewis, S., Hughes, P. R., *Chem. Commun.* (1965) 489.
(61) Stern, E. W., Spector, M. L., *J. Org. Chem.* (1966) **31**, 596.
(62) Paulik, F. E., Roth, J. F., *Chem. Commun.* (1968) 1578.
(63) Evans, D., Osborn, J. A., Wilkinson, G., *J. Chem. Soc. (A)* (1968) 3133.
(64) Slaugh, L. H., Mullineaux, R. D., *J. Organometal. Chem.* (1968) **13**, 469.
(65) Blum, J., Rosenman, H., Bergmann, E. D., *J. Org. Chem.* (1968) **33**, 1928.
(66) Tsuji, J., Ohno, K., *J. Am. Chem. Soc.* (1966) **88**, 3452.
(67) Baird, M. C., Nyman, C. J., Wilkinson, G., *J. Chem. Soc. (A)* (1968) 348.
(68) Ohno, K., Tsuji, J., *J. Am. Chem. Soc.* (1968) **90**, 99.
(69) Alderson, T., Jenner, E. L., Lindsey, Jr., R. V., *J. Am. Chem. Soc.* (1965) **87**, 5638.
(70) van Gemert, J. T., Wilkinson, P. R., *J. Phys. Chem.* (1964) **68**, 645.
(71) Cramer, R., *J. Am. Chem. Soc.* (1967) **89**, 1633.

(72) Kagawa, T., Inoue, Y., Hashimoto, H., *Bull. Chem. Soc. Japan* (1970) **43,** 1250.
(73) Malatesta, L., Santarella, G., Vallarino, L., Zingales, F., *Angew. Chem.* (1960) **72,** 34.
(74) Maitlis, P. M., Pollock, D., Games, M. L., Pryde, W. J., *Can. J. Chem.* (1965) **43,** 470.
(75) Smutny, E. J., Chung, H., Dewhirst, K. C., Keim, W., Shryne, T. M., Thyret, H. E., *Am. Chem. Soc., Div. Petrol. Chem. Preprints* (1969) **14** (2) B100.
(76) Stern, E. W., Spector, M. L., Leftin, H. P., *J. Catalysis* (1966) **6,** 152.
(77) Kohll, C. F., Van Helden, R., *Rec. Trav. Chim.* (1968) **87,** 481.
(78) Volger, H. C., *Rec. Trav. Chim.* (1968) **87,** 501.
(79) Stern, E. W., *Trans. N.Y. Acad. Sci.* (1970) **32,** 66.
(80) Smidt, J., Hafner, W., Jira, R., Sedlmeier, J., Sabel, A., *Angew. Chem.* (1962) **74,** 93.
(81) Clark, D., Hayden, P., British Patent **1,111,714** (1968).
(82) Bailey, G. C., *Catalysis Rev.* (1969) **3,** 37.
(83) Zuech, E. A., Hughes, W. B., Kubicek, D. H., Kittleman, E. T., *J. Am. Chem. Soc.* (1970) **92,** 528.
(84) Calderon, N., Ofstead, E. A., Ward, J. P., Judy, W. A., Scott, K. W., *J. Am. Chem. Soc.* (1968) **90,** 4133.
(85) Lyons, J. E., *Chem. Commun.* (1969) 564.
(86) Dent, A. L., Kokes, R. J., *J. Am. Chem. Soc.* (1970) **92,** 1092.
(87) Jardine, I., McQuillin, F. J., *Tetrahedron Letters* (1966) 4871.
(88) Garnett, J. L., Kenyon, R. S., *Chem. Commun.* (1970) 698.
(89) Schrauzer, G. N., *Accounts Chem. Res.* (1968) **1,** 97.
(90) Murray, R., Smith, D. C., *Coordination Chem. Rev.* (1968) **3,** 429.
(91) Allen, A. D., Bottomley, F., *Accounts Chem. Res.* (1968) **1,** 360.
(92) Cramer, R., Lindsey, R. V., *J. Am. Chem. Soc.* (1966) **88,** 3435.
(93) Hirai, H., Sawai, H., Ochiai, E., Makishima, S., *J. Catalysis* (1970) **17,** 119.
(94) Moiseev, I. I., *Am. Chem. Soc. Div. Petrol. Chem., Preprints* (1969) **14** (2), B49.
(95) Hopton, J. D., Swan, C. J., Trim, D. L., ADVAN. CHEM. SER. (1968) **75,** 216.
(96) Cullis, C. F., Trim, D. L., *Discussions Faraday Soc.* (1968) **46,** 144.
(97) Clark, D., Hayden, P., Smith, R. D., *Am. Chem. Soc. Div. Petrol. Chem., Preprints* (1969) **14** (2), B10.
(98) Clark, D., Hayden, P., Walsh, W. D., Jones, W. E., British Patent **964,001** (1964).
(99) Kitching, W., Rappoport, Z., Winstein, S., Young, W. G., *J. Am. Chem. Soc.* (1966) **88,** 2054.
(100) Schultz, R. G., Gross, D. E., ADVAN. CHEM. SER. (1968) **70,** 97.
(101) Tamura, M., Yasui, T., *Chem. Commun.* (1968) 1209.
(102) Van Helden, R., Kohll, C. F., Medema, M., Verberg, G., Jonkhoff, T., *Rec. Trav. Chim.* (1968) **87,** 961.
(103) Bryant, D. R., McKeon, J. E., Ream, B. C., *Tetrahedron Letters* (1968) 3371.
(104) Clemment, W. H., Selwitz, C. M., *Tetrahedron Letters* (1962) 1081.
(105) Gow, A. S., Heinemann, H., *J. Phys. Chem.* (1960) **64,** 1574.
(106) Knapsack, A. G., German Patent **1,252,662** (1968).
(107) Robinson, K. K., Paulik, F. E., Hershman, A., Roth, J. F., *J. Catalysis* (1969) **15,** 245.
(108) Report of the Organometallic Chemistry Panel, Science Research Council of Great Britain, Nov. 1968.

RECEIVED September 16, 1970.

12

The Role of Oil in the Year 2000 in Canada

D. K. McIVOR and A. W. BROWN

Imperial Oil, Ltd., 500 Sixth Ave., South West, Calgary 1, Alberta, Canada

In Canada as in other countries whose economy is an advanced state of development, growth in the energy consumption has closely paralleled growth in the economy, but the share of the energy market supplied by various fuels has changed dramatically. Economic growth, energy consumption growth, and fuel shares have been extrapolated to the year 2000. The results indicate that Canada's oil consumption rate will triple over the next 30 years and that the demand for energy usage will be so high that petrochemical feedstocks will be a scarce commodity.

Almost anyone who can extrapolate trends can make forecasts. Forecasts generally only assume significance for those who examine them when:

(1) a rationale for their general credibility is developed.

(2) their significance to real-world events is estimated.

Thus, most of these remarks are structured around the credibility of this forecast of oil's role 30 years hence and its significance to our society.

The quantities presented are intended to be directional only. Anyone who has examined the results of past long term forecasts is aware that they are seldom highly accurate. The hazards of prophecy are such that "good" long term forecasts prove to be directionally correct but almost never highly accurate, while "poor" ones are neither; the best that can be hoped for is that this forecast will be proved correct in the direction and the order of magnitude of the changes it suggests. It should also be noted that in many areas the forecast raises more questions than it provides answers.

Discussion is restricted to the role of oil in the year 2000 in Canada, with external factors receiving notice only insofar as they affect the

Canadian scene. This does not reflect insularity as much as lack of sufficient background to address the topic on a wider basis. Although many areas in which the forecast has considerable significance to the oil industry and Canadian society have been identified, it has been possible to deduce only one major significant point relative to the chemicals industry.

The Forecast

Thirty years ago in 1940 coal was Canada's chief fuel, supplying well over half of the country's energy needs. Although it is probably difficult for many young people to credit, wood was a widely used fuel. Electricity was, of course, widely available, but it was expensive. Gas supplied about 3% of our energy, oil less than one-quarter. Nuclear energy did not exist. Three events generally referred to as revolutionary have affected our energy consumption patterns in the past 30 years. Two of them, nuclear energy and the jet engine, resulted from World War II. The first resulted in a new energy source, the second created a large new demand for a traditional source. The third event, the dieselization of railroads, was like the second inasmuch as it created a relatively rapid growth in demand for an existing source.

It is worth examining the actual meaning of the adjective "revolutionary" which is generally used in connection with these three events. As to nuclear energy, the first full-scale nuclear electrical generation station began operation in 1956, and 14 years later this source of energy supplied only a very minute fraction of the world's electrical generating capacity. The first jet powered aircraft flew during World War II, but the first commercial jet flight took place seven years after the war's end in 1952, and widespread commercial use did not come about until 1959. Dieselization of Canadian railroads was phased-in over a period of eight to nine years. Thus, in the frame of reference of energy consumption patterns, a revolutionary event has meant a new supply or demand which causes or portends a significant alteration but which in fact requires almost a decade or more to take effect.

The more profound effects of the past have been trends which have developed over even longer periods. Perhaps the most important trend or relationship is the growth in the energy consumption of nations which accompanies economic development. It might be possible to formulate an axiom which says, "a nation can possess energy supplies but lack significant economic development, but there can be no significant economic development without energy supplies."

Figure 1 illustrates the strong correlation between national economic development and energy consumption. An attempt to normalize the scale effect has been made by presenting the data on a per capita basis. Thus

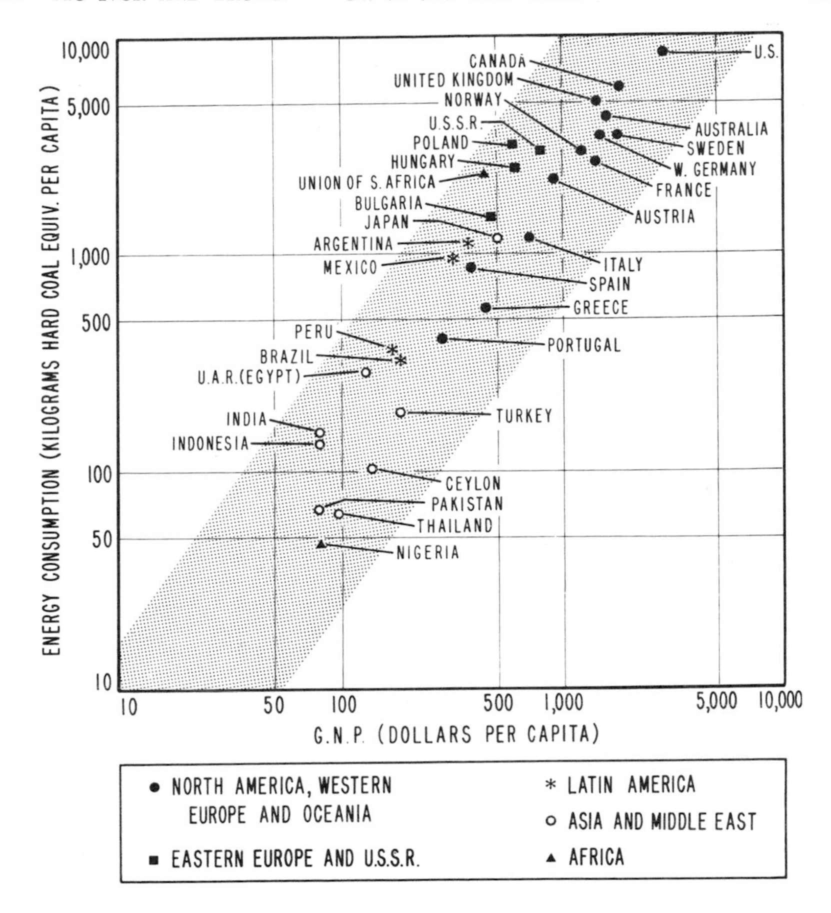

Figure 1. Energy use vs. *GNP*

while the U.S.S.R. has a greater population than the United States, its per capita energy consumption and gross national product are significantly less; even more dramatic is the difference in this regard between Canada and India. The scales on this chart are exponential so that the disparity between nations at either end of the curve is very large. Figure 1 of course captures the economic development and energy consumption of several nations at a point in time. Obviously growth in the first must be accompanied by growth in the second.

Figure 2 illustrates that except for the anomalies related to the World War II, Canada's economic growth as measured by real (or uninflated) gross domestic product has been accompanied by nearly parallel growth in energy consumption.

Figure 3 illustrates the combined effects of evolutionary trends and the so-called revolutionary events. (It is smoother than Figure 2 simply because it has been plotted at five-year rather than one-year intervals.) In 1940 coal and wood together supplied 69% of Canada's energy, and by 1970 this was reduced to 15%. Hydroelectricity grew apace with energy demand until most generating sites within economic transportation distance of markets were developed. Then, of course, its growth rate slowed. Nuclear energy consumption is still too small to be visible on this chart, amounting to 0.1% of the total in 1970. Its growth has been slower than most earlier expectations mainly because of very high capital costs and because technological evolution has not been as rapid as expected.

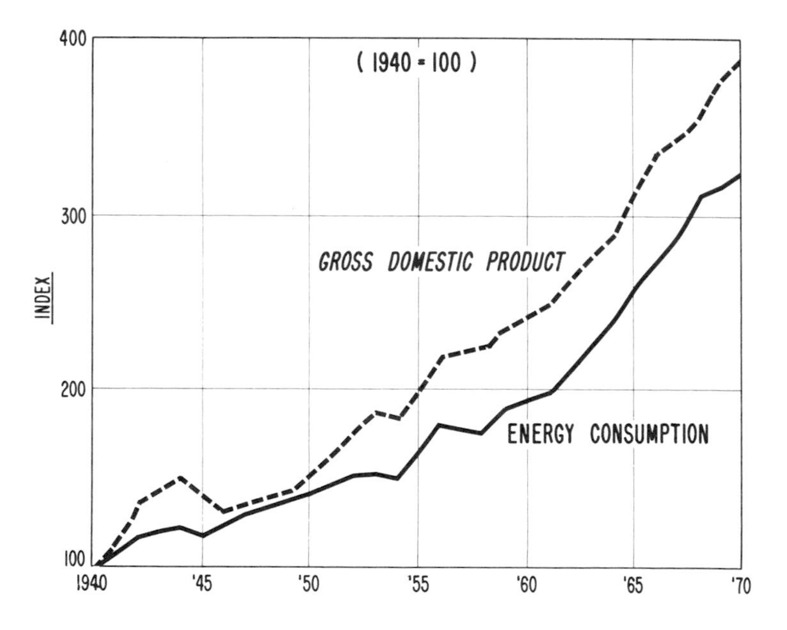

Figure 2. Economic growth and energy consumption in Canada (1940–1970)

Obviously most of the past 30 years' growth has been captured by oil and gas. Oil's availability, price, and portability has made it the fuel which drives virtually all transportation in a nation which depends abnormally on transportation; its price, convenience, and cleanliness has allowed it to take almost half of the industrial and residential/commercial heating market. Natural gas has seen a phenomenal rise during the period as a supply for the heating market. The major determinant in this increase at this particular time was related to the fact that Canadian supplies large enough to warrant transport to distant large markets became available only relatively recently. This volume availability has, in turn, been largely a by-product of successful Canadian oil exploration.

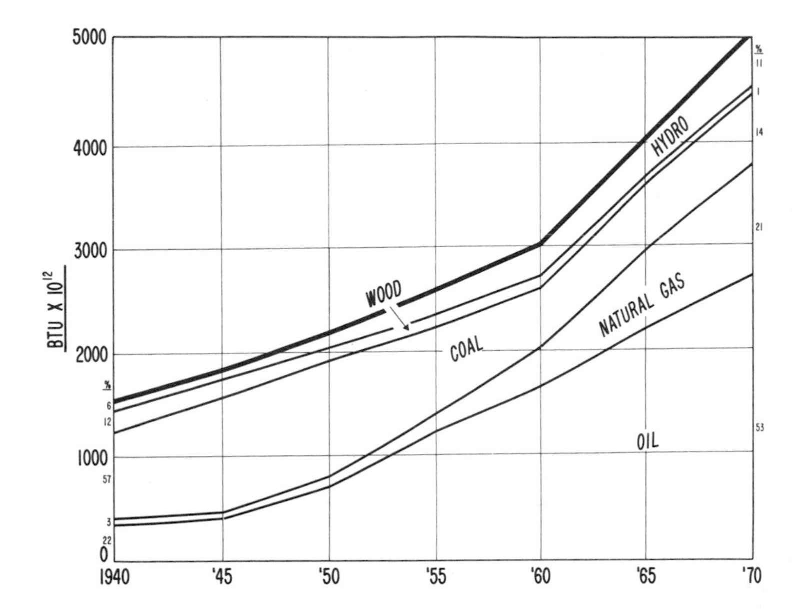

Figure 3. Energy demand in Canada (1940–1970)

With all these factors as inherent reasons, the Canadian energy market is over three times as large in 1970 as it was in 1940 and is today supplied 53% by oil, 21% by natural gas, 15% by coal and wood, 11% by hydroelectricity, and 0.1% by nuclear-derived electricity. The changes in either total energy consumption or the shares supplied by competing fuels have occurred gradually even with the events categorized as revolutionary included.

At last the major questions are posed: what will Canada's energy consumption be in the year 2000, and how will this demand be shared among fuels? On the question of how much energy Canada will consume 30 years hence, considerable discussion has been given to developing the close historical relationship between economic growth and energy consumption. Thus, the first step in looking for an answer to this question lies in speculation about Canada's future economic growth. When the controversy generated by the recent five-year economic forecast by the Economic Council of Canada is considered, it is quite natural that a 30-year forecast is approached with some trepidation.

Canada has now had 25 years of almost uninterrupted economic growth. In other words, there has been neither a major economic depression arising from internal domestic causes nor one developed in sympathy with continental or world events. Neither has there been an international armed conflict of sufficient proportion to cause world, continental, or Canadian economic depression; of course, it is generally accepted that any future major international conflict will involve nuclear weapons with consequences so terrible that there would be small probability of recovery. Thus, the forecast should be approached with the assumption that the next 30 years will contain neither major international conflict nor major economic depression since the effect of the first would probably be an end to our society as we know it, and the effect of the second is essentially unpredictable. The tactic to adopt is to assume that Canada's economy will continue to grow at approximately the trend rate established during the past 25 years and then to speculate as to whether such a growth could possibly be accomplished.

Actually, a slightly more sophisticated approach has been taken. For the years 1970 to 1985, the economic forecast is based in some detail on the relations between economic growth and growth in population, human productivity, and capital productivity, and projections of these determinants. Beyond 1985, however, the 1985 "real" gross domestic product has been simply extrapolated at a constant compound growth rate of 4.7%. The result, shown in Figure 4, indicates that in the year 2000 Canada's gross domestic product could be about $215 billion relative to today's $55 billion. (These numbers are in constant or uninflated dollars, related for statistical convenience to the year 1961.)

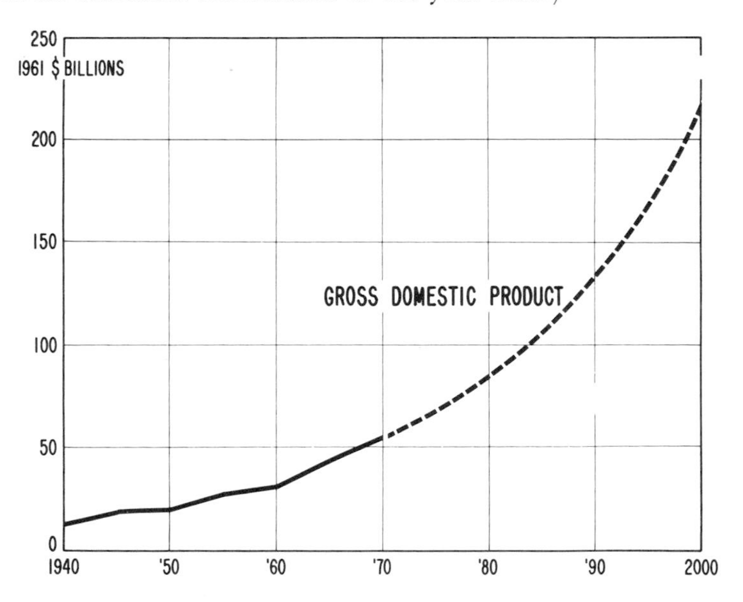

Figure 4. Economic growth in Canada (1940–2000)

Can such economic growth actually be achieved? Many specific questions suggest themselves. For instance, how much of the growth would be achieved simply by additions to the labor force and investment capital, and how much might arise from increased human and capital productivity? If productivity is constant at 1970 rates, Canada would need about 75 million people to support a gross domestic product of $215 billion. Many people would probably doubt that Canada's resources would support 75 million people sharing increasingly better economic circumstances.

It seems much more logical to assume that productivity will continue to increase. The question can be approached in another way by adopting a year 2000 population forecast of 35 million or one which trends in fertility, mortality, and immigration suggest. On this basis, gross domestic product per capita would be $6200 or almost two and a half times 1970's $2600. Will technological advances allow Canadians to raise their productivity this much in 30 years? Canadian productivity was doubled in the last 30 years, so the forecast seems to be within the realm of possibility. As another perspective, competent economists predict that in the United States a figure of $6000 per person will be reached as early as 1985.

A second question concerns how such economic growth would be managed. The present and future governments of Canada would have to include economic growth at this approximate rate among their objectives and continue to foster a climate in which it is possible. Speculation on this question could be and probably will be the subject of scores of papers more comprehensive than this. Therefore, it is only noted here that important determinants will be government's philosophy toward the proper share of national wealth to be used by government to promote social and political objectives, the degree to which inhibition of inflation affects economic growth, and relationships with the United States.

Speculation on the credibility of this forecast of economic growth is ended here, and the reader is asked to accept it as being within the bounds of possibility. In translating it into energy consumption, the tactic of assuming that historical relationships will prevail is again adopted, and speculation as to whether this is reasonable is entertained.

Figure 5 shows the result. If the historical relationship between economic growth and energy consumption continues and if the economic forecast is approximately correct, then by the year 2000 Canada will be consuming over 20 quadrillion Btu's compared with slightly over 5 quadrillion in 1970. This is less than one-third of United States' consumption today. On a per capita basis, it would mean that for each of the 35 million Canadians assumed at that time, about 585 million Btu's would be consumed annually compared with about 250 million today. (Again the competent economists quoted earlier suggest that per capita consumption in the United States will reach 450 million Btu's as early as 1985.)

However, will Canadians actually increase their consumption of energy about two and one-half times in the next 30 years? If so, how? In the last 30 years our per capita consumption has increased 75% for these major reasons: we travel a great deal more, each of us occupies more heated space in the winter, energy consumed by industries has grown faster than population, and we make our lives easier or fuller by a fantastic array of conveniences and gadgets not available 30 years ago.

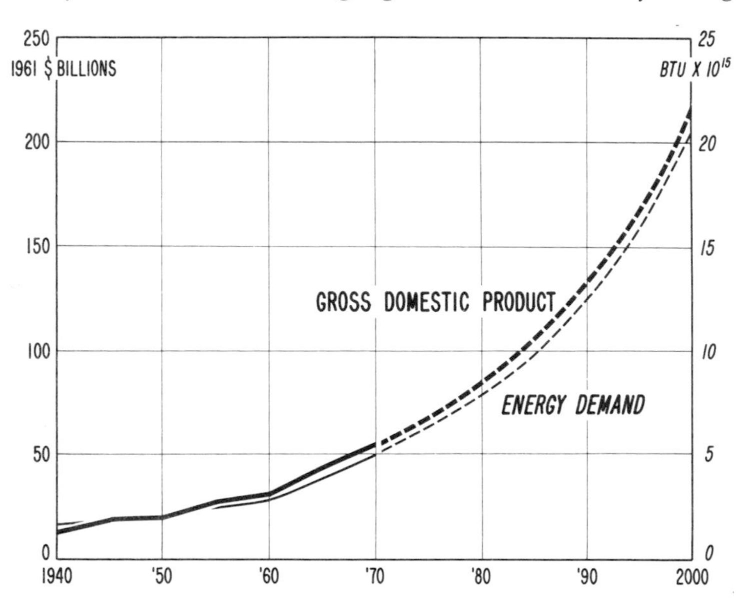

Figure 5. Economic growth and energy demand in Canada (1940–2000)

Over the next 30 years the per capita use of energy could be increased two and one-half times if the changes in our lives which appear likely are considered. For instance, since it seems inevitable that urbanization will continue and our megalopolae will continue to grow, the expansion in public and private urban transportation will create large energy demands. So will the far-from-saturated market for long distance travel. Increasing service and recreational activities will generate a direct demand and an indirect one through their economic multiplier effect. In addition to new demand created by growth in traditional industries, there will probably be new demands relative to industries created to process our expanding waste materials and to preserve or reclaim the condition of our surroundings. Finally, the development of the Canadian north has inherent large transportation demands for people, for inbound materials and supplies, and for outbound products. If this increased energy consumption is to occur, it must be within the constraints imposed by controls on air, water, and other pollution.

Again speculation on the question is dropped, and the reader is asked to accept that it is within the bounds of rational speculation that Canada will consume over 20 quadrillion Btu's of energy in the year 2000. The second major question posed earlier is: how will this approximate demand be split among competing fuels? Between 1970 and 1985 a considerably detailed forecast was made using an energy consumption model developed by the Imperial Oil Co. This model estimates the energy required in major "end use" categories by correlating demand with several selected economic indices. Fuel shares within each end use are estimated by extrapolating historic growth/decay trends. The projections are finally modified to reflect specific expected events, but essentially the estimates are projections of historical trends.

Beyond 1985 extrapolation of the growth/decay trends of competing fuels has simply been continued, with the only modification being related to the assumption that new natural gas supplies will be available in Quebec and the Atlantic Provinces, either arising from discovery in the coastal sedimentary basins or through imported liquified natural gas.

Figure 6 illustrates the result. Natural gas and nuclear energy are indicated to increase their market share substantially. Water power declines from 11% to 5%, oil from 53% to 45%, coal and wood together from 15% to 8%.

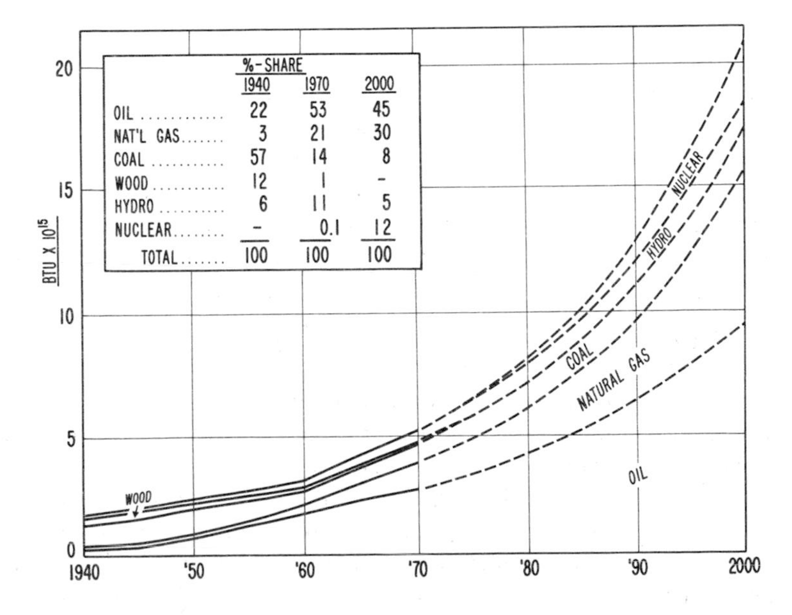

Figure 6. Energy demand in Canada (1940–2000)

Since this estimated share pattern was derived mainly from projection of trends (particularly long term trends), it seems appropriate to focus on oil and speculate as to how possible future events might alter its forecast future role. Events related to pollution control tend to indicate increases in petroleum demand. The use of lead free gasoline, for instance, requires additional refinery processing, which in turn consumes more petroleum fuel. Increasingly tighter controls on sulfur dioxide emissions from thermal-electric plants will cause a shift from coal to low sulfur fuel oil if there is no economic flue-gas desulfurization to cope with coal's sulfur content.

A large increase in market share for natural gas has been indicated. Natural gas is in relatively short supply and is significantly different from oil inasmuch as its offshore import is today expensive. Thus, fulfillment of this share will require large new indigenous supplies and/or significant decline in offshore transportation costs. A large increase in consumption of nuclear energy has been shown, despite difficulties currently being encountered. Faster or slower nuclear growth will be reflected in oil's share, primarily as a determinant of the amount of oil consumed as fuel for electric generation. In the transportation market, it is difficult to visualize a change which does not continue to consume petroleum products. Electric cars would create a tremendous demand for electricity for battery recharge and might cause a shift rather than a decline in petroleum usage. Steam cars still need fuel to generate steam, and it is hard to see a fuel whose portability or convenience surpasses oil products.

In summary, the forecast for oil's small decline in market share over the next 30 years is a net decline resulting from a smaller share for coal and larger shares for natural gas and nuclear energy. Events which could accentuate the decline are greater availability of nuclear energy or gas or the development of an as yet unforeseen transportation method. On the other hand, the decline would be slowed or compensated by lesser availability in either nuclear energy or gas. Events related to pollution control indicate greater petroleum usage and a shift away from coal.

Perhaps the most important point of all is that *even* if oil's share is assumed to decline 8% over the next 30 years, in absolute terms there will still be a tremendous increase in Canadian oil consumption. As indicated in Figure 7 this would be about 5 million barrels per day compared with about 1.5 million per day in 1970. (It has been assumed that the demand for petrochemical feedstocks will increase seven times in 30 years. This reflects the assumption that the high growth rates of the formative years of the Canadian petrochemical industry will not be sustained in the future. Nevertheless, the assumed growth is faster than projections for chemicals in general and faster than the GNP growth.)

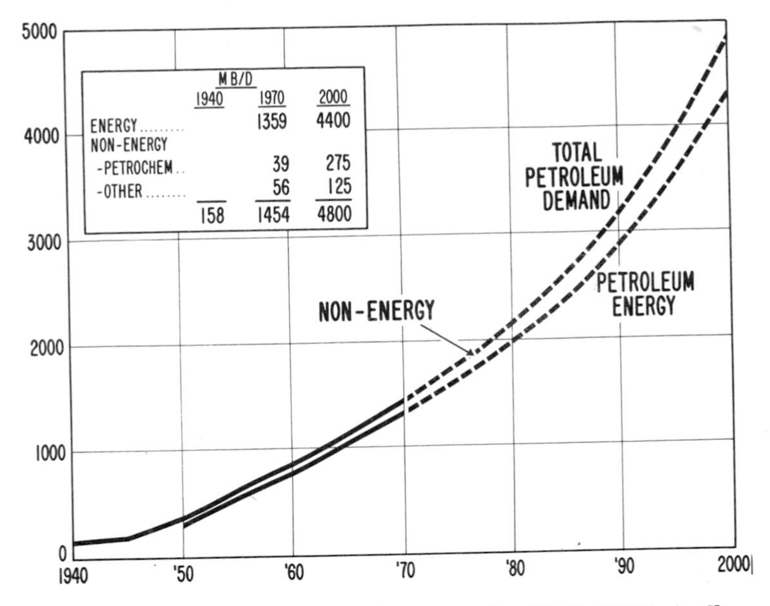

Figure 7. Petroleum demand in Canada (1940–2000) (million barrels per day)

Implications

The growth in petroleum consumption will axiomatically be concentrated if the trend toward urbanization continues. Obviously city and municipal planners will have their work cut out for them in designing urban centers compatible with the amount of local and distant transportation implied (roads, airports, public transportation). Industrial planners will probably find it necessary to design smaller cars. Canadian petroleum refining and marketing capacity will have to expand four times, and these facilities must be compatible with urban plans. Oil industry planners will be challenged similarly to marshal resources (men, capital, equipment, and materials) to supply both the petroleum raw materials and the distribution, refining, and marketing system required. Perhaps the most critical requirement will be for capital because, as others have recently highlighted, this will be very large. Most of the remainder of this paper develops this subject, starting out with the premise that 30 years hence Canada will be consuming on the order of 5 million barrels per day of oil.

Starting with the production phase, we assume that if the domestic supply is sufficient and properly geographically located, Canada will fill its own domestic needs for oil. If Canada is presumed to approach self-sufficiency in the early 1980's, between now and the year 2000 the industry would need to find about 35 billion barrels to end the century with a reserves/production ratio of 10 years. This is about twice the average finding rate of the past 20 years but may well be possible because since 1965 exploration in Canada has moved to the several as yet largely unexplored sedimentary basins which fringe Canada's periphery: the Arctic Islands, the Beaufort Basin surrounding the Mackenzie Delta, the Pacific Offshore, the Atlantic Offshore, and Hudson's Bay. The Canadian Petroleum Association has estimated that the crude oil potential of Canadian sedimentary basins is on the order of 120 billion barrels. While this type of estimate can be considered only directional at this stage of exploration in the frontier sedimentary basins, if the real potential is only half as large as the estimate, Canada will be able to supply its domestic needs and also have substantial exports. In regard to long term petroleum supply, Canada is extremely fortunate in also having large proved reserves of heavy hydrocarbons or tar sands whose production is technologically feasible.

Going back to the 35 billion barrels which would be required for domestic needs over the next 30 years, the possible magnitude of expenditures needed to locate, develop, and produce these should be examined. It must be presumed that much of the new oil will be located in frontier areas remote from market and thus will have to be found and developed much more cheaply than the Western Canada reserve to be profitable. On this basis, a notional estimate of $25 billion is suggested.

Domestic petroleum product requirements of 5 million barrels per day would mean that about 3.7 million barrels of refining and marketing capacity would have to be put in place over the next 30 years at a cost of about $6.5 billion, not including replacement or maintenance investments for refineries now in place. Domestic oil transportation costs are difficult to estimate, but it seems reasonable to add at least $4 billion which would perhaps cover the ultimate costs of two major new pipelines plus unspecified tankers and other facilities. Thus, a staggering total of perhaps $35 billion can be visualized to meet domestic petroleum needs alone. The word "staggering" is not too strong when one considers that in the 23 years since the discovery of Leduc, only $17.5 billion has been expended by the oil industry, including investments for natural gas. This is equivalent to the entire private capital investment for all causes in Canada for the past three years or is equivalent to almost half of Canada's 1969 gross national product expressed in 1969 dollars.

Recent comments by others indicate that this forecast does not single out Canada as unique. For instance, Frank C. Osment, President of the Standard Oil Co. of Indiana, made these points during a recent speech before the Organization of Petroleum Exporting Countries (*1*). His comments relate to the next 20 years rather than 30 years:

World petroleum consumption will rise from 40 to 100 million barrels per day.

Supply of this demand will require that the petroleum industry find and develop 615 billion barrels of new oil.

The costs associated with finding, developing, producing, transporting, refining, and marketing this oil will be on the order of $650 billion over the period.

Conclusions

Over the next 30 years, Canada's economy, expressed on the basis of gross domestic product in 1961 dollars could grow from $55 billion to about $215 billion. This growth could be achieved by population and productivity increases which are possible as judged by historical Canadian increases and shorter term projections of future U.S. economic performance. If energy consumption continues to bear approximately the same relationship to economic growth as it has in the past, Canadian energy consumption will rise from 5 quadrillion to 20 quadrillion Btu's per year. Fuel shares within this total demand have been projected on the basis of a history that includes evolutionary trends and so-called revolutionary events. These projected shares for the year 2000 appear reasonably consistent with trends and events which speculation indicates could occur over the next 30 years. While oil's share of the year 2000 total Canadian energy market declines, its growth in absolute terms is large. There is basis for considerable optimism that Canada will be able to supply its own oil requirements as well as substantial exports.

The capital requirement for finding, developing, producing, transporting, refining, and marketing the required volumes is large compared with historical requirements. It would about equal the entire private capital investment which has taken place in Canada over the past three years. Comments by others regarding the same factors worldwide, however, indicate that Canada is not unique in this requirement for large petroleum investments.

Finally, only one general conclusion has been identified which has a bearing on the chemical industry. While oil supply in any given country rarely even approximately matches oil demand at any given time, it appears that in Canada over the next 30 years the demand for energy purposes will be large. Therefore, it seems unrealistic to expect that Canadian oil will be a glut on the market and thus be available at bargain prices for petrochemical feed.

Literature Cited

(1) Osment, Frank C., *World Oil* (Feb. 1, 1970), 61.

RECEIVED May 26, 1970.

INDEX

INDEX

I

K

L

M

N

O

P